DIVIDED

A WALK ON THE CONTINENTAL DIVIDE TRAIL

BRIAN CORNELL

ISBN 978-1-6957337-5-6

Cover Design: Alpha Vision
Book Design: Lyubomyr Yatsyk
Cartographer: Jordan Cummings
Author Photograph: Masashi Matsubuchi
Terminus Photograph: Green Kat Photography

For…

Those who have and will again.

Those who have and can no more.

Those who have not but one day will.

CONTENTS

Part Four: Idaho & Southern Montana

Part Five: Northern Montana

AUTHOR'S NOTE

This book describes my experience hiking the Continental Divide Trail. The days, hours, and miles recorded here were recalled through personal memory, photos, videos, and journals. Most names (in addition to trail names) have been changed to protect the privacy of hikers, helpers, homeless, and townies alike. All views and opinions expressed are my own.

INTRODUCTION

Six months ago, my older brother, Nick, moved back to Mammoth Lakes, California. About an hour's drive south from Yosemite National Park, this magnificent mountain town sits a snowball's throw from the awe-inspiring Sierra Nevada mountains. Having spent the past four winters and two summers here myself, living the ski bum life, I began to desire a new challenge. Soon after my brother's arrival, he informed me of his plans to hike the Continental Divide Trail, beginning in April of 2018.

The Continental Divide National Scenic Trail stretches from the floor of New Mexico to the crown of Montana. As you can deduce from the name, it follows the corridor of the Continental Divide vertically across the United States. With two possible start points, two acceptable endpoints, and many alternates, routes, roads, and paths along the way, the distance of this trail can vary anywhere from 2,400 to 3,100 miles. It meanders through three national parks, one national monument, twenty wilderness areas, and five states: New Mexico, Colorado, Wyoming, Idaho, and Montana.

Nick tossed out an invitation to me as if it were a tri-fold pamphlet. He was going whether I said yes or no. Completing this trail would grant him what is known in the hiking community as the Triple Crown. This achievement entails hiking all three major long-distance National Scenic Trails in the States. In addition to the CDT, this trifecta includes the Appalachian Trail and the Pacific Crest Trail.

It would be my second long trail, having already hiked the Appalachian Trail with him in 2014. That metaphorical pamphlet planted a seed that quickly grew. Within the next month, I began telling myself, *Yes, go hike!* Soon after, I began telling others, "I'm hiking the CDT this summer!" Once speaking this aloud, there was no turning back.

Six months later, and here we are. My gear is dialed, and I have $10,000 in the bank. Backpacker Law states, "While thru-hiking, you will spend around $1,000 each month, give or take a few hundred depending on your frugality or extravagance." I did some research on alternates, towns, resupplies, and mileage, but wanted as much as possible to be left for the moment. The unknown is a beautiful part of the journey. Too much research and portions of the hike would be spoiled. I have never set foot on any part of the Continental Divide Trail, so all of this hike will be through unfamiliar territory.

Part One: New Mexico

I.

ONCE A THRU-HIKER, ALWAYS A THRU-HIKER

April 30 - Day 1

It is just another early morning at the Econo Lodge in Lordsburg, New Mexico. I am not sure what other clientele this hotel supports, but hikers seem to be their main source of income during this season. Some of us fill our stomachs at the continental breakfast before leaving. We slowly make it out to the parking lot. It's still dark outside, and a few stars remain shining in the western edge of the sky. Ten hikers and two drivers mill around a pair of large rental trucks. Tired faces hold half-cooled cups of coffee or full bottles of water.

The Continental Divide Trail Coalition runs a shuttle service from Lordsburg to the southern terminus of the trail: the United States/Mexico border. From there, we will hike back up to Lordsburg for a resupply before continuing north.

It seems silly to be taking a vehicle down to the border just to walk through the same town we left a few days prior, but I am not going to get started on what's silly. Not yet.

I put my backpack in one of the large black trash bags provided and set it down in the bed of the truck. It's a dry ride to the border, and you do not want to start off the hike with a dust-covered backpack weighing you down and wearing out any material it infiltrates. There will be plenty of time to get yourself and your belongings covered in dirt over the next few weeks, so there's no need to get a head start. We pile into the trucks, three across the back and two up front with the driver.

The CDT Coalition rents full size, four-door trucks, so there is adequate space for six grown adults. They rent these vehicles from early March to mid-May, when most northbounders begin their journey. Any earlier than this and you arrive in Colorado before the winter's snow has a chance to melt. Any later and you are hiking through the New Mexico desert in extreme heat, with water sources dangerously low or else completely dried up. Our driver is a volunteer and well-rehearsed in navigating down to the border. The thru-hiking season and shuttles having started nearly two months earlier, he is plenty comfortable getting to the drop-off point and leaving hikers stranded in the middle of the desert with only their backpacks and each other.

We get rolling down the asphalt road and introduce ourselves. Leah is upfront in the middle, the driver, Joe, and my brother, Nick, on either side of her. She is the only female amongst the nine male hikers being taken down to the border this morning, a standard ratio. Across the back sits Adam, Will, and myself. We also provide our trail names, but only three of us have them: Will is Tarzan, Nick is Nacho, and I am Knots.

A trail name is usually given to a hiker early on in their hiking career. It is essentially a nickname or a moniker to adopt in the woods to mask your true identity. We leave a society we know so well to walk in the backcountry for half the year, becoming outsiders to the world that recognizes us by our original name. Going on trail is a rebirth of the self and one that requires a new tag to establish your commitment to life as a thru-hiker. A trail name is usually out of your control. It depends on who you hike with, who you speak to, and what sticks to your aura as a hiker.

For the first fifteen minutes of the ride, Leah is on the phone with her car insurance company, having waited until the last morning to cancel her automobile coverage. Today, we are all making the transition from human to hiker, choosing to leave our roles in society to become a wanderer of barren lands through the Continental Divide passageway.

As hikers, we are a leech of civilization, spending limited time in the small towns that line the trail, a passive sightseer never to be seen 'round these parts again, and a tourist in the rawest sense of the word. These are kind, somewhat romanticized descriptions, yet this is the desired life for a thru-hiker and the type of lifestyle that constantly haunts us between hikes.

We are going as far south as we can along the Continental Divide in the United States, where the Divide is virtually non-existent, hidden beneath ripples of desert sand. We turn off the highway and onto a dirt road. The sun rises to our left as anticipation grows with the day. Nerves in my stomach are tossed around as the truck rocks back and forth down the bumpy road. Joe sits comfortably behind the wheel, navigating over the battered desert with ease. Seven weeks into the hiking season, and he well knows every foot of dust that makes up the dirt path to the border. Windows remain up during the ride. The black trash bags in the bed have taken on a layer of desert camouflage.

We take three stops for pee breaks. We are all relieved to be back in Nature, free to pee wherever, especially in the desert, where we don't have to worry about urinating too close to water sources or streams that aren't there. All passengers are clearly well hydrated and prepared to walk in eighty-degree sunshine across the exposed desert. Grateful for the fresh air, we breathe deeply and stretch our limbs. I'm never one to get car sick, but this ride is not an easy roll down the highway.

Included in our shuttle package is the placement and upkeep of five water cache lockers between the southern terminus and Lordsburg. We drive along a low mountain range as Joe informs us where the water caches are located, tucked behind desert brush and hidden out of sight from the road. The first eighty-six miles of trail have but a few reliable natural sources, and these caches save us from having to haul an absurd amount of water between them.

On the other side of the truck, Tarzan begins to eat an apple from the hotel. He devours the whole thing in primal fashion, stripping the scarlet ball down to a palmful of brown specks. Hikers do not like to waste food. We don't like to waste anything, actually. For most of us, this isn't our first long-distance hike, and we are all properly trained to Leave No Trace: the ethical principles carried by most hikers and travelers of the outdoors. These habits are well ingrained in our hiker code, having been learned and developed on prior backpacking trips.

There isn't much conversation amid the bouncing around inside the cab. I am glad to be wearing a seatbelt in this full vehicle with bodies that provide complementary cushion. I grab onto Nick's headrest in front of me as I try to keep a level field of vision. I figure I might as well get used to looking at the back of his head.

Nick completed the Pacific Crest Trail two summers ago, in 2016, and during the summer of 2014, we hiked the entire Appalachian Trail. From Georgia to Maine, we camped every night together and always hiked within a few miles of one another. We took no days off amidst the 2,185-mile journey. I did my best to keep up while doing my part to slow him down as much as possible. After 131 days of playing hiker tug of war, we summited Katahdin, capping off our thru-hike of the Appalachian Trail.

Four hours after leaving the Econo Lodge parking lot, we arrive at the bottom of New Mexico. The southern terminus is marked with a granite obelisk, an information board, and a small pavilion, providing respite from the sun. Ten yards away, the international border is marked by a crude barb wire fence. Anything more would leave a scar on this landscape. No more than four feet high, this primitive combination of wire and wood reinforces the fact that the landscape itself deters most from pursuing the American Dream.

Tarzan walks to the border and stares out over the fence.

"You watering the plants over there?" I call out to him.

"Yeah, they're lookin' a little dry!" he yells over his

shoulder, still facing Mexico. He pees through the fence, not in an attempt to disrespect our southerly neighbors, but in a sentimental manner to honor his excitement of being here and the beginning of his journey.

We all take pictures with the monument, marking the start of our new lives as hikers of the Continental Divide Trail. Today marks the beginning of a 2,500-plus-mile journey, from the bottom of the country to the top, and a day that marks the commencement of many good days to come. It is only the beginning, yet I tell myself that getting to this point was the hard part. Working two jobs through winter was the hard part. Saying goodbye to friends, which damaged some relationships while strengthening others, was the hard part. Leaving a town that became home was the hard part. There is difficulty in saying goodbye, but I knew once I left, the next five to six months would hold nothing but bliss and tranquility. Doubts murmuring in the back of my mind fade away to be replaced by words of encouragement and thoughts of elation.

I have an idea of what to expect in terms of lifestyle and demands, not only physically, but mentally as well. Those of us who have thru-hiked know the feelings that accompany such a journey and yearn to experience these emotions again. We have felt the overwhelming joy of a thru-hike and, conversely, have suffered the lowest lows that break us down into tears of pain and shouts of frustration, with no one around to listen or help. You can scream at the hills in front of you, cry to the sky, or else plead with the trees for the pain to stop, but it never does. The only solution is to breathe and walk.

No matter how low you get while hiking, the highs are worth every step. There will be bad moments during this journey, but the worst day on trail is still better than the best day off it. We have made the sacrifices necessary to be here and must face the miles that await. As thru-hikers, we are chasing the high of being able to walk all day, through the longest days of the year and the hottest hours of the season. We watch the days get longer and hang on to every minute as they slowly shorten. We desire to accomplish something most only have the energy to scoff at while feeding off their looks of uncertainty and words of desire. We yearn to complete this challenge, to achieve something measurable by days and miles but enjoyed more for the things that hold no measure.

Time between hikes is spent reminiscing over our months and miles in the wilderness as a backpacker: an unemployed tramp of the trail, an evolved caveman tuning into the energy of Earth and returning home. Knowing this life exists is a curse. It haunts your thoughts while you work, encourages your identity off-trail, fuels your heart to love, and promotes the pursuit of simplicity in all aspects of life. Regardless of how simply I live off the trail, I believe the only truly simple life is one spent outside, smelling sage, and sweating under the sun while carrying everything you need on your back. We wander like the children of Neverland, wishing it was Everland. Looking for an "X" on the treasure map, we search for elusive riches. Riding a sea of unknowns, endless opportunities await.

The sun stands high in the sky, and after taking enough pictures, it is finally time to start hiking. Goosebumps cover my body, and I pinch myself, giddy with excitement. Smiling dumbly, my nervous legs carry me over desert sand and between needled cacti. Tears of joy escape my eyes as the smile on my face breaks even wider. We are actually on the Continental Divide Trail. All that is left is to walk.

II.

COUNTING CROSSINGS

May 10 - Day 11

I step to the edge of the river and look across. Some of the crossings have cairns to indicate where the trail goes, but this one is not so obvious. I prefer to hike without my glasses on, but the elusive trail along the Gila River encourages me to wear them. In the heat, they slip down my sweaty nose and require a return to the bridge that often leads to scuffed lenses. Even with optical assistance, I find it difficult to spot the continuation of trail from here. It'll be easier to find once I cross the water.

Upon entering the knee-deep river, I see fish dart about my legs. They scatter from the feet infiltrating their home, yet curiosity brings them closer. Smaller fish remain near the protection of rocks while larger ones wiggle around the deeper pools. The water is cool and refreshing, perfect for a hot spring day in New Mexico. My feet have been wet since 6:30 this morning, so there is no point in trying to rock hop over the river. Even if it can be done at one junction, it is not possible at all of the two hundred other crossings. This is crossing 115 for the day, and it is only 3:00 p.m.

The trail crosses the river quite frequently on this alternate, so I count out of curiosity and to give me something to do. Yesterday, we only crossed it forty-nine times, but we also stopped for a few hours at Doc Campbell's Post, where we picked up pre-sent resupply boxes of food. Even with a three-hour break at Doc's, during which we rested, charged devices, and ate two lunches, we still managed to hike twenty miles.

The Gila River alternate is one of the routes spoken highly of by those who have hiked the CDT in previous years. The Gila Wilderness was the first designated wilderness area in the United States and is the largest wilderness area in New Mexico. The "official" CDT loops around to the east, but the trail through the Gila is shorter and has much more water availability. The past week and a half in the desert was hot, with distances between water sources increasing.

Halfway through the river, I take out my empty bottle, unscrew the filter, and dip it into the water. A few seconds later, the bottle is full. I screw the Sawyer squeeze filter back on and bring it to my mouth. The flow rate is not great, but it's the most convenient and easy-to-use water filter that I know of. I suck and squeeze until the bottle is half empty. I unscrew the top, refill, and return it to my mouth. In the river, rushing water is the only thing to be heard. During the day, the soft background noise sets my mind at ease while reminding me to drink plenty of fluids. There is no need to look ahead to see where the springs are or carry any water between them. Crossings are so frequent that I only have to stop and drink whenever I am thirsty.

I reach around to stuff the empty bottle back into the side pocket of my pack before splashing my face with more water. For the first time all trail, I have felt properly hydrated for multiple days in a row. Not having to worry about conserving water, I am able to drink my fill until my urine comes out clear the other end. After walking through the dry desert for over a week, we are thankful for the excess in water, even if it means having soaking wet feet for four days straight. This is day three walking along the Gila, and the scenery around every turn of the river is well worth soggy feet and stage one trench foot.

My legs are red from shin scratching bushes and, undoubtedly, covered with traces of poison ivy. I take advantage of the many river crossings to rinse my legs and relieve them of an incessant itching. On the other side of the river, I walk across a stretch of sand that gives way with every step. I force myself through more overhanging bush to find the trail once more.

The overgrown trail causes some discomfort but, overall, I am grateful to be walking through this luscious canyon. There is much more shade, and at times, it feels like a jungle. The thick forest softens the sound of the Gila River. Wildflowers bloom in patches, and tree branches hang low overhead. It's a pleasant change of pace from the dry, brown desert we were walking through for the first week.

The trail is easy enough to follow once I actually find it, but the river crossings have been so frequent that my pace is slower than usual today. Some trail is decently tracked, but other sections are seemingly nonexistent. I come upon another crossing: number 116. I pause halfway across to remove my glasses and cup a few more handfuls of water to my face. The late afternoon sun remains above the canyon, but the river does well to cool me off.

Sand on the side of the river sticks to the bottom of my shoes. The combination of sand and water makes my feet heavy, which is another reason for the slower travel today. With every step through the soft banks, my shoes gain slightly more weight. I spot a cairn and head toward a part in the grass to the trail beyond. A stack of rocks or a sequence of footprints in the sand is all I need to locate the path. My eyes are learning to be more adept at recognizing trail, though sometimes, I am left to navigate where there appears to be no established route, hence the plethora of cuts on my legs from periodic bushwhacking. The river is flowing south, so as long as I walk the opposite way, I know I am headed in the right direction.

Crossing 117 comes quicker than expected, but I stomp through nonetheless. The trail mimics the river and turns a corner. The tight canyon of the past few miles opens up to a spectacular view of sheer rock walls towering high, hundreds of feet above the river. Spires stand tall, carved by the water and wind. The sun is temporarily blocked from view as shade stretches across the canyon. My eyes are grateful for the shadows. Luckily, the sun is only visible between the tall canyon walls from around 10:00 in the morning until roughly 4:00 in the evening. The miles before and after this window are much easier than those that fall inside of it.

I walk slower in the shade, enjoying a darker environment before the river undoubtedly turns again to receive the sun in full force. Here, the river widens with the canyon as the water consequently shallows. Crossing number 118 is only ankle-deep, and crossing 119 brings me out of the shade, back into the warmth of the sun. It is not a direct route, walking with the water, but around every corner, there is another surprise.

The river twists back and forth, following a well-established path. After spending many lifetimes carving out the landscape, water rolls along with little intention as evidence of time and Nature's subtle force.

At the next crossing, I see a snake slide into the water as I approach the edge of the river. I interrupted its sunbathing, and now I'm about to interrupt its water bathing. Snakes used to freak me out, but this is the third one I have seen in eleven days, so I'm getting quite used to them. I don't want anything to do with them, and it is obvious they feel the same way about me. I look into the water to make sure the snake isn't where I'm about to place my feet. Crossing 120 is a little deeper, this one rising to my mid-thigh on the far side.

From the little research I did, I knew there would be a lot of crossings. I repeat the number in my head between crossings to not lose track of the count. This gives me something to focus on and is a distraction from the miles. As the day wears on, my mind tires of keeping track, but I continue to do so. I cannot be upset with the water, it is just there and has brought me so much joy, comfort, and refreshment the past few days.

We are planning to do twenty-four miles today. Nacho and Tarzan are somewhere ahead of me. I have not seen either of them since we took our lunch break around noon. My brother is usually in front, and I have given up trying to keep pace with him. A few days ago, I got the motivation to try but was discouraged by his effortless overtaking of me after a couple of hours. A hiker such as Nacho is tough to slow down, and this is going to be our longest day yet. The miles go by like a blur, and I worry that at this pace, not only will my body be unable to keep up, but my mind will not be allowed adequate time to appreciate the environment and to investigate the corners that demand attention.

The desire to be somewhere else and constantly chasing what's next is a demon many travelers must attend to. No matter where you are, there is always another place you have not been that is waiting to be discovered. As a hiker, you are moving more often than not, but this movement must be done carefully to appreciate all that is around you while maintaining the motivation to keep moving in order to see more, not moving too slowly so that progress is inhibited but not too quickly so that everything remains unnoticed even after passing through.

Constantly, we are reminded to enjoy the fleeting beauty as it slowly passes by and to live in the present. If I walk too quickly or attempt too many miles, my eyes are often looking down at the ground, and I struggle to witness the beauty surrounding me. Every morning, I remind myself we are still walking through New Mexico. From the low desert to the high desert, and all of the canyons, rivers, and rolling grasslands between, our surroundings are ever-changing.

Each day has become increasingly more frustrating as I keep up with Nick. I arrive at camp after him, exhausted from the day's walk, just to see him seated comfortably, well-rested, and full of energy, ready to hike another five miles. For me, the enjoyment of hiking comes from being out in Nature, while his enjoyment comes from seeing how many miles he can do in a day, pushing his body to the limit, only to increase those limits even further.

When we were planning out a rough itinerary, we estimated we would be averaging twenty miles a day through New Mexico. We barely kept to that for the first week and are now averaging slightly more. Hiking twenty miles a day is enough of a challenge, and any more causes me pain, stress, and irritation: pain from my body not being in shape yet for twenty plus mile days, stress from trying to keep up with my brother, and irritation from ignoring the scenery and focusing on miles more than the moment.

I do not mind hiking alone. In fact, I prefer it. While my legs are not in their best shape, they are only getting stronger by the day. The first couple of weeks on trail are usually used as training for most hikers. After a few hundred miles, you will become a desert crossing and mountain climbing machine. Your legs will be stronger than ever, though your upper body will lose most of its muscle. Walking twenty miles a day will soon be normal, and day hikers will be passed with astonishing ease. It is possible to hike or run before leaving for a thru-hike, but it's difficult to do fifteen to twenty miles a day when you are still working a full-time job and leading a social life. The only way to get in shape for a thru-hike is to go on a thru-hike.

The next crossing is more of a river walk. The trail on the other side is indiscernible, and it will probably just cross again a little further along. I turn left and walk up the river. My wet feet stay in the water as I pull them through against the current. The canyon is narrow, and the trail will probably continue to cross back and forth for a while. Constantly looking for the trail is exhausting and mentally tiring. I relish the break and am glad to walk in the river. The fish swim swiftly out of my way as I disturb the water's usual progress.

After a hundred yards in the river, I exit onto the bank and search for the elusive markings of a trail. The sound of tumbling water grows louder, and around the next turn, I am greeted with a small cascade pouring down the canyon. The falling water drips along, one plateau of rock at a time, as the sound echoes against the tight walls. All sounds within this narrow corridor are louder than they would be if left to open air.

The river is the only thing to be heard as I step closer, drowning out the sound of any whistling birds, overhead planes, or approaching hikers. At the bottom of this small cascade is a large pool. The sun is still just above the canyon walls, sparkling against the whitewash. I take my pack off, remove my shirt, and wade out into the water.

It is just the perfect temperature. My body soaks up the water, and I completely submerge myself, combing the water through my hair as I blow bubbles to the surface. The sound of the cascade is muffled, and the pool of water provides an escape from the heat of the late afternoon. I resurface, release a primal yell, and shake out my hair. Standing with eyes closed and arms open to the sun, I allow the warm ball of light to dry me out as I inhale deeply on the shore of the river. Water is life, and life is grand.

III.

INTO PIE, OUTTA TOWN

May 14 - Day 15

Every three to seven days, we either hitchhike to or walk through a town where we are able to stock up on food to last us until the next town. Pie Town, our next resupply, is still about fifteen miles away. The sun slowly approaches the horizon. This whole day has been hot and exposed and my water supply is getting low. The map claims a water source a few miles before the highway, so I begin to drink what I have left, planning to refill at this next tank before finding a campsite for the night.

I smoke an evening bowl and suck down the rest of my water. I see a solar panel off in the distance and begin walking a little faster toward it, empty bottle in hand. I see Nacho and Tarzan already there, hiding in the shade of the solar panel, presumably filling up, hydrating, and waiting for me.

Nacho is smiling suspiciously, and Tarzan has his phone plugged into an outlet that is attached to the panel.

"How much water do you have?" Nacho asks.

I hold up my empty bottle.

"Anything in your other bottle?"

"Nope," I reply.

"You drank all your water?"

"I was thirsty."

"There's no water here," he continues.

"Really?"

"Really."

"You can charge your phone though," Tarzan informs me, waving his phone in his hand.

"Any service?" I ask.

"Nope."

"How much water do you guys have?"

"One liter," answers Nacho.

"Two liters," replies Tarzan.

Upon rechecking the map app, recent comments offered up by earlier hikers admit to this being a dry source. We are laughing the whole time, giggling at our stupidity for overlooking this one important detail. Our thirst blinded us. The sun deteriorated our minds. There is no other water source on the map between this point and Pie Town, so we throw out our best ideas. We decide the only thing we can do is what we always end up doing anyway: keep walking.

Instead of walking straight toward the road, we go off route and blaze a diagonal path across the farmland to increase our chances of finding another water tank elsewhere amongst the grazing land. While hiking through New Mexico, we try to be cognizant of whose land we walk on as the trail often borders private land or follows a road that allows easement. Some landowners are overprotective of their acreage and claim their cattle become stressed when strangers wearing backpacks walk past. Landowners themselves can get upset, and I am sure they all have more guns than they can carry. This is in the back of our minds, but no one speaks it aloud as we make our way across whoever's property we're on.

Our course is set on the highway, but we change direction when we see another water tank in the distance. Crossing under a barb wire fence, we work our way toward this marker. Hopes are extinguished when we look over the edge of the tank and find it to be dry, as well. We will have to keep walking until we get to the road or find water, even if we have to suck it from a cactus or knock on a stranger's door to ask for it. Thankfully, we are all well hydrated, so missing a water source is not too detrimental to our health at this point.

A mile further along, we come upon an enclosure that has a couple of large tires within it. Thick with algae, one of these tires has just enough water to dip a bottle into. Looking closer, I notice clusters of cow hair floating on the water and chunks of algae drifting below the surface. This is what water filters are for. We sit and drink while the evening sun continues to heat us up. With spirits slightly higher and water bottles filled with a slurry mixture of cow hair and algae, we keep walking toward the road still on the lookout for a healthier-looking water tank.

The next one we reach is overflowing. The tank is shoulder height, and the pipe coming out of the ground fills a large cylinder as it spills over the sides, returning to the dirt it just escaped from. The ground is wet, and we step on strategically placed rocks to reach the pipe and fill our water bottles. We are obviously not the first thirsty souls to take advantage of this particular windmill. A "No Trespassing" sign, posted on the structure, is ignored.

Spirits are even higher after departing this proper water source and having our whistles thoroughly wettened. We continue toward the road and figure we can camp close to it, out of sight from passing cars and potential landowners. Walking side by side, we allow plenty of room between us to minimize our damage to the desert plants. When walking off-trail through the desert or a meadow, it's common practice to spread out, so you don't walk the same route and trample on one plant or area too many times. A small bush or patch of grass can recover a lot quicker if only one foot steps on it as opposed to three.

Plenty tired from our exciting evening, we reach the road, worn out but grateful to be well hydrated and carrying full water bottles. Small trees beside the road offer protection from the highway while the forgiving desert sand comforts our backs. Even after patching a few holes in it the other night, my sleeping pad is still leaking. Any soft ground to sleep on is a bonus, not that I should have any trouble falling asleep after an exhausting twenty-eight-mile day. We go to sleep dreaming of pie, resolving to rise early to walk the eight road miles into town before the asphalt has a chance to heat up too much.

***

May 15 - Day 16

The Toaster House is easy to find amidst the rundown trailer homes and shacks. The fence is lined with toasters of all varieties, shapes, sizes, and colors. Some have flags and pictures painted on their sides, while others are signed with kind messages. There are at least ten other hikers at the house when we arrive, which is quite overwhelming at first. This is the largest congregation of hikers we have seen since Silver City, a week and a half prior. Hikers tend to gather in town and usually stay at the same hostel or cheap motel. This hostel is the only place to stay in town, so there really is nowhere else to go. If it was not for the Toaster House, Pie Town would have been an easy town to pass through without allowing a second thought. Or a second slice.

There are a couple of tents set up in the yard, clothes hang from the line, and hikers mill about the front porch, warming up for the day. We drop our packs on the ground and approach curiously, introducing ourselves to the hikers we have not met yet while reading the signs and notices posted on the walls. After skimming through the house rules and other random tidbits of information, I get the gist of the hostel. Treat the place with respect, don't leave your stuff lying around, clean up after yourself, and please donate so the Toaster House can be enjoyed by others for plenty more years to come. There is a donation box in the kitchen for hikers to drop money in, so Tina, the owner, can pay for electricity, water, and other general upkeep of her old house.

The hostel has bunks in two of the rooms, beds in another, and a couple of couches in the living area. The hiker box is overflowing with every hiker food imaginable, and dishes are stacked neatly in the drying rack. There is one bathroom with a shower and a washer, but the toilet is in the backyard, consisting of two outhouses, one of which is a double-seater. Resupply boxes are stacked in the dining room. I make sure mine is there, but don't remove it, not wanting to sift through its contents just yet.

There are not only hikers staying here but a handful of bikers as well. Some of the hikers at this place have already been here for a few days, and we reconnect with Eighty-Five, Leah, Nightwalker, and Spider (who all started the same day as us) while meeting plenty of other new hikers. It is still early on in the trail, and we are all fairly close to one another. Some are gathering their belongings and getting ready to leave, while the rest sit around enjoying the shade and conversation.

The rest of the morning is spent talking with other hikers, doing laundry, and resting. Hikers are a good bunch to talk to, and conversation varies wildly across the front porch. Spliffs are rolled, and bowls are smoked, as we take advantage of the shaded seating while enjoying the company of smiling faces. Two hikers are discussing Eckhart Tolle while another is talking loudly to no one in particular about how he left Alabama and fell in love with hiking. Tina shows up in the late morning and shares some information about her town, including the fable of how Pie Town got its post office.

Many years ago, soon after Pie Town was established, the then-owner wrote to the Postmaster General, asking for a post office to be built in their town. The general's reply claimed the United States Postal Service would not establish a branch in a place called Pie Town, and they should change the name of the town before asking again. The proud owner wrote back, telling them the town's name is Pie Town, and if they don't like it, they can take their post office and go to hell. I guess the Postmaster General preferred the cool confines of his air-conditioned office because he ordered a branch to be built, and it is still called Pie Town to this day.

The Gatherin' Place is the only restaurant open during the week, and soon enough, we are all hungry again. It's a mile down the road, so naturally, we walk there, happy to sweat a little bit and stretch our legs. Three long tables sit in the large open dining space, a chalkboard on the wall displays the day's menu, and a tall glass cupboard containing pie stands along the wall. They have all the flavors you could ever desire and then some. There is apple, butterscotch, blueberry, raspberry, blueberry raspberry, mixed berry, cherry, chocolate cherry, chocolate pecan, pecan, pecan custard, pear, pear ginger, peach, and their famous New Mexico apple pie which is made with green chili and pinyon nuts. I marvel at the selection before taking a seat and ordering a mess of home fries with a slice of apple.

The waitress takes our orders back to the kitchen, which is visible from the dining room. The cook is busy at the stove, cracking eggs and flipping flapjacks. He wears a cowboy hat and a handlebar mustache.

"Put some music on, wuddya?" he asks the waitress.

She nods and returns to the dining area, "Hey Alexa, play Waylon Jennings."

"Mama's Don't Let Your Babies Grow Up to Be Cowboys" begins to play from a hidden speaker in the corner of the dining room. The waitress pulls pie out of the ten-foot-tall cupboard, serving them with an unhealthy amount of ice cream and Reddi-whip.

There is a good mixture of locals, tourists, and hikers intermingled at the three tables. Locals wear ten-gallon Stetsons while they eat and catch up on recent town gossip. Tourists goggle at the immense pie selection, having only been drawn here because of the town's name amid a road trip. Hikers talk to all while enjoying the air-conditioning that is far more refreshing than a stale desert breeze. The restaurant is aptly named, as residents of Pie Town surely have nothing better to do than to shoot gossip and eat at the only place open in town.

Regulars come in and have their usual: a slice of pie and a cup of coffee. The doughnut has no place in such an establishment. Obviously, the only way to wake up in New Mexico is with a strong cup of coffee and a triangular confectionary. Locals have gotten used to seeing hikers come through and are happy to share their cafe and pie with us. Food is served, and we are granted with portions large enough to fill even a hiker as the waitress is kept busy refilling coffee mugs and water cups. I feel spoiled drinking clean water without having to suck it through a filter or worrying about where my next refill will be.

We return to the hostel well-fed, exhausted, and grateful for a shorter day. I am glad to have some time off my feet and out of the sun. The past few days were tense between Nick and me. He kept suggesting we hike a thirty-mile day, and I kept refusing. "Wanna do a thirty today, Knots?" His tone is jovial and sarcastic, but I know he would be doing thirty-mile days if I wasn't with him. "Hell no," I would answer. Hardly two weeks in and my body is still adjusting to the demand of hiking every day, hardly considering doing any more mileage than we already are. The twenty-eight miles yesterday was our longest day so far but also an accident, thanks to our miscalculation of water sources. It wasn't the worst thing, but we only did it out of necessity.

We talk while at the Toaster House, and Nick voices the idea of attempting the Fastest Known Time (FKT) on the

Colorado Trail. The Colorado Trail goes from Denver to Durango, sharing four hundred miles with the Continental Divide Trail along the way. He joked about doing this during a couple of our research sessions, but the seed was planted way back then, and now, it is forcing its way to the surface.

I am not too surprised at this crazy declaration of his, as if hiking the CDT wasn't crazy enough already. This being his third long trail, he strode past crazy a long time ago. At this point, it's normal to be hiking all summer, and it's normal for him to be able to hike thirty-mile days back to back to back. It is also an excuse to hike faster and leave me behind. He asks if I want to do it too, but knows my answer before the words leave his lips. After some research and calculations, he figures he will start to increase mileage out of Grants, New Mexico, which is our next resupply point.

It is clear we want different things out of this hike and have our own motivations for being out here. Nick is able to push himself in certain ways to achieve his goals. For me, the satisfaction I get from doing big miles is not equal to the dissatisfaction I feel from having not appreciated the bit of trail I just rushed through. His motivation is to hike as far as possible each day, pushing his own body to limits that are constantly developing and entering new thresholds. My motivation for hiking this trail is to take a vacation and spend five to six months in the wilderness. The faster you hike, the sooner it is over. I don't ever want this to be over. If you are always looking ahead to where you want to go next, it is difficult to appreciate where you are right now. The moment is impossible to find when you are too focused on what is coming.

The AT proved to us we can hike a long trail together, and it was nice to have my brother there to share that experience with. This time around, I know better and understand there is no point in hiking a hike that isn't my own. Nick is also hiking a hike that isn't his own. I am fine doing twenty miles a day, but he would rather be doing thirties every day. Instead of doing what we each want to do, we compromise somewhere between. It becomes increasingly more frustrating to be pulled farther than I desire to go, while he feels restricted in his own progress.

There is no right way to hike a hike, just as there is no wrong way to hike a hike. Backpacker Law states, "Hike your own hike." There are no rules about how many miles you have to hike a day, where you have to start and end, or how you are supposed to travel between towns. The official route is no more than a red line on the map, and there are plenty of ways to get from point A to point B.

For Nick, setting the goal of attempting an FKT on the Colorado Trail gives him something to work toward. He will increase mileage after Grants and walk all the way to Denver before turning around to begin his attempt going south, from Denver to Durango. This will give him plenty of time to get in even better shape while allowing him a look at what the Colorado Trail is like. He has already hiked from Mexico to Canada on the PCT, and knows he can do this, so why not throw in an out-and-back of the Colorado Trail to add another challenge to this already difficult hike? I am slightly disappointed and somewhat sad that my brother will be walking away from me come next town, but also feel a separate excitement growing in myself to be on my own on a long trail for the first time.

***

May 16 - Day 17

Ate a lotta pie:
Peach, apple, blackberry pie!
Breakfast and dessert.

We enjoy one more meal at The Gatherin' Place before returning to the hostel to dump the contents of our resupply boxes into our respective food bags. By 2:30 in the afternoon, we walk out of Pie Town. A few miles outside of town, cars drive by intermittently, blowing up the dust. It glitters in the sun before settling back down on the road. Not much else to do in Pie Town, might as well go for a drive. One guy even buzzes by on an ATV. Most of the people slow down, wave, and give us plenty of room as they pass.

Nacho and Tarzan are a few hundred yards ahead of me. I watch a truck slow down and stop for a moment to talk to them. They wave it off, and it continues in my direction. The driver comes to a stop beside me and holds up a water bottle.

"Ya' need any?" he asks.

I am plenty thirsty in the afternoon heat and keen on taking a short rest.

"Sure, why not?" I say, really just wanting to talk to one more Pie Towner before leaving that town to the dust. I lean in on the passenger window, take a water bottle from him, and turn my back to the sun. "What are you up to this evening? Was that you that went by on the ATV?" I ask.

"Yeah, I came out to get a buddy's truck started and am bringing it back to town for him."

The ATV is in the bed of the truck. I gulp down water while he talks and swallow a few mouthfuls before responding.

"Looks like you got her goin'."

"Wasn' too difficult," he admits, shrugging his shoulders. "Y'all walkin' that trail?"

"Yup. What do you know about it?" I ask, before drinking some more water.

"I know y'all are crazy," he says, laughing. I see he is missing a handful of teeth. He takes a sip of the Budweiser he's holding in his left hand.

"We're all a little crazy." If we weren't crazy, we'd miss out on a whole lot. I tilt my head back and drain the water bottle.

"We used ta call y'all footbackers. Walkin' the same trail as them guys that come through on horses."

"Footbackers?"

"They ride horses, so they're horsebackers. Y'all on your feet, so we call ya' footbackers. Now I guess you're jus' hikers."

He hands me another full bottle. I unscrew it and dump it into my empty bottle.

"What's your name, man?" I ask, holding out my fist for a hiker handshake.

"Mark," he replies, reaching out a hand before doing a double-take at the sight of my knuckles. "Wanna smoke?" He reaches into his door pocket and brings out a small pipe. It's fully packed and lightly toasted.

"Yeah, alright. Thanks."

I take a hit and hold it out for him.

"That's all you, man," he says, taking another swig of Bud.

I smoke some more of his bud, turning the green flower to ash.

"Pie Town is an interesting place. I think you're a little crazy for living here, but what more do ya need?"

"Could do with a few more women," he admits.

We both laugh. Good point. I did not see many women in Pie Town, let alone attractive women. Maybe they were right to get out of town.

"I hear ya," I reply, looking around to all the females not on trail beside me. "Say, what's with these cows down here? They get all in a fuss whenever we walk by them."

"Yeah, they aren' too fond of you hikers. They're use' to seein' people on horses so when they see y'all on two feet an' wearin' a backpack, they get a little spooked."

I thank him again for the water and weed before continuing north. The other two are only getting further ahead, and I have to catch up with them at some point.

A COMPARISON OF COWS AND HIKERS

Despite the obvious differences, of how many legs we choose to walk on, and the fact that hikers are not shot in the head when they get old or fat enough, we have a lot in common. The similarities are numerous, and we cross paths many times throughout the day.

We walk the same land. Wandering around under the hot sun all day, we trot down dirt roads and often find ourselves on the same stretch of trail. Cow paths are spider-webbed across the desert and can be useful when traveling cross country, beside a road, or to and fro water sources. These paths are well worn in, as their hooves stomp the ground to a fine powder. Many times, I have come across a cow either on trail or right beside it. They hurry away as I approach, looking back at their pursuant while trotting ahead. Cows standing off to the side extend looks of curiosity toward the intruder walking by on two legs.

We drink the same water. This valuable resource is scarce in the desert, and usually, the source is a big rubber tire or large steel tub. The water comes from a well or, if we're lucky, a piped spring from somewhere deep in the hills. Sources tapping the deep water table are commonly powered by the sun or wind, both of which are plentiful in the desert. Windmills are easily spotted at a distance and are more common than solar panels. Solar panels provide a little shade and are slightly more up to date than the ancient windmills, some of which look to be a hundred years old. They stand tall and creak with every gust of wind that blows against their rusted blades. No matter the vehicle for energy, these pumps pull liquid gold right out of the ground for which both the cows and I are indulgent of and grateful for.

These water sources are popular hangout sites for hikers and cows alike. They are a great place to take a long lunch or late afternoon snack, but remember, cows can get defensive of their water and do not appreciate intruders. Watch out for the big, fat, ugly bull kicking its leg, and be ready for a false charge. They may not understand English, but talking to them as you approach will do well to calm your nerves and uneasiness.

We shit on the same ground. Granted, I dig a cat hole every time and pack out my used toilet paper, while the cows leave behind shit pies a foot in diameter to bake in the sun, five times larger than any of the craps I have ever laid in my lifetime: large, dry craps that I toss aside to clear space for a campsite in the evenings. These cows obviously do nothing but eat all day, judging by the size and multitude of these brown discs lying about the desert.

We sleep on the same dirt. Only having used my tent a handful of times, I spend the majority of my nights cowboy camping in the desert. They call it "cowboy camping" because it's how the cowboys used to sleep when out in the fields and under the stars: jacket for a pillow and hat over the eyes. Or maybe it's called cowboy camping because we are just boys out there camping with the cows, keeping them company. Or possibly because anyone who is willing to chase cows around all day should be considered a boy.

Some may say, "Hey Knots, aren't you worried that a cow might step on you or a snake might slither into your quilt in the middle of the night?"

Fair question, to which I retort, "Can't worry if you're not awake!"

Many nights I have been gently mooed to sleep by a chorus of cows as they slowly step past in search of a quality sleeping spot themselves (although come to think of it, I have never seen a cow sleeping). Before falling asleep, I yell to my four-legged friends to let them know I am lying nearby, politely asking them to not step on me as they go wandering about in the middle of the night. The cows are now my only friends in the desert, and their grunts fill the otherwise empty space. They do not just *moo,* and it is unfair to assume any of Old McDonald's animals are only capable of producing one sound. They *merrrrr* and *ereighhh* but also *mmuuaaaahhhh* and *uherrerrhh.* They are more reliable than the rooster for a wake-up call, but do not expect a straight answer when asking for directions.

Another difference between cows and hikers is that cows cannot cross cattle guards or fit through fence stiles. They are restricted to a certain acreage. This keeps them confined to one space, whereas I am truly free, allowed to continue north as I please.

IV.

SAVED BY THE SHADE

May 24 - Day 25

It has been another gorgeous day in the desert. Before starting the CDT, I had never spent much time hiking in a desert environment. My unfamiliarity with this ecosystem encouraged me to tighten my hold on the mountainous territory I am so fond of. I grew up in the mountains and feel at home when amongst them. Their high peaks provide protection, and their bountiful lakes supply adequate fuel. My discomfort with desert terrain eventually led me to face it, and I have become a more well-rounded hiker because of it.

Surprisingly, I have thoroughly enjoyed hiking through the expansive desert of New Mexico. The open landscape is laid out in front of me, and the flat miles make for easy hiking. The sun's arrival and departure cause the sky to glow, projecting colors seen only in the sky. The ground is alive, crawling with lizards, snakes, and spiders. My vitamin D levels are overflowing, and water is appreciated more when it is so scarce. A little more careful planning is required, but after encountering the dry source outside of Pie Town, I have not miscalculated since. Hiking alone out here has made me slightly more cautious. I have to make all decisions independently, and if something goes wrong, I have only myself to blame. On the contrary, I have not fucked up yet, so my self-esteem and confidence in my solo hiking abilities are higher than they have ever been.

Exposure in the desert can be draining at times, but I do carry an umbrella, which I pop open whenever it is flat and not windy. When taking breaks, I make sure to do so in the shade. Consequently, as there is not much shade, I am walking for extended durations between breaks and taking longer rests to take advantage of the shade when it is available. I break under short trees and behind large rocks. The sun moves shadow across the ground, forcing me to move with it. I do my best to plan out my days so lunchtime siestas take place at a water source, but this is difficult to ensure since there are so few water sources already. Most of the time, I deal with the heat and am used to it by now. The desert is no longer frightening and unfamiliar to me.

At every source, I drink heavily before refilling my four-liter capacity. My level of hydration is measured by the color of my urine and frequency of pee breaks. Walking has become rhythmic by now, and my trail legs have come along nicely. In just over a week, I will be entering Colorado and facing the mountains that make up most of the trail throughout the state. The desert has been relatively flat, and while I can now hike twenty miles comfortably, this is different walking when compared to the aggressive terrain of Colorado.

During the heat of the afternoon, I come upon a short tree and a tall rock that looks to have tumbled down from the nearby mesa many years ago. There is just enough space for one hiker and a backpack. The rock provides a cool backrest as I sit and snack in the shade. I have been alone for six days now and am enjoying every bit of it. The voice in the back of my head that's usually saying, *Hey B, Nick is waiting on you*, is silent, and so are my days.

Most days, I do not see anyone. The solitude is appreciated, and my mind is at ease. I have not seen another thru-hiker in a few days, and it is quite odd to go a whole day without seeing or talking to anybody else. The only ones walking trail out here are CDT hikers, and I have not seen a weekender or day hiker since the Gila Wilderness. Most people choose to walk somewhere a little more protected and slightly more frequented by water.

Behind the rock, I check the map app and consider the water situation. There is a water cache coming up at the next road crossing, and I am relying on this supply to get me through to the next natural source. Even though Backpacker Law states, "Never rely on a water cache," I choose to temporarily ignore this certain proclamation. The water cache is between a particularly long dry stretch, and it would be thirty miles from one source to the next without it. Those who tend to the water cache know this, which is why they have stockpiled water here in the first place. Being right by the road, it is easily accessible, albeit a decent drive from town. The trail angels who take care of this water cache assist every hiker that comes through this section with such crucially placed water jugs.

I hear footsteps approaching and wait until they pass me before making my presence known. A young man walks by in salmon-colored shorts and a navy blue button-up. It's Eighty-Five, a fellow hiker who started the same day as Nick and me.

"Hey, Eighty-Five!" I shout, standing up to be seen.

He jumps, as expected, and turns around.

"Hey, Knots! Didn't see you there."

"Yeah, I was hiding in the shade. How have you been? Still hiking with Leah?"

"A little hot but can't complain. Yeah, she's somewhere behind me. Where's Nacho?"

"He wanted to hike bigger miles, so he's probably a couple of days ahead at this point. Let me pack up real quick. I haven't seen anyone in a while."

I gather my things and walk with him to get some conversation in for the day. He hikes fast, and I'm moving quicker than usual to keep up with him. Soon, I fall back and let him carry on his pace without me. I'll see him at the water cache in a few miles anyway.

When I reach the road, my mouth is incredibly dry, and my body is overheating. It was a hot haul to this road crossing, and I have been conserving my water. You know, just in case. I don't see Eighty-Five anywhere and observe no sign of a water cache. I continue across the road and walk a little farther up trail to see him sitting underneath a small tree with thirty water jugs tied up, tucked beneath the shade. He is sitting beside the water jugs and eating what looks to be a cookie. There is also a small cooler here filled with snacks.

"Yesssss!!" I yell to the blue sky as I drop my backpack in a cloud of dust. I retreat under the shade, pull out a water bottle, and begin drinking.

Free snacks are great, but the only thing I care about now is water. A few of the one-gallon jugs are empty, but most remain full, ready to be drunk. A notebook lies open on Eighty-Five's legs. Messages and notes from previous visitors fill the lines of the dust-stained sheets. I recognize signatures from those ahead of me, including Nacho and Tarzan, who were here two days ago, with my brother including "Hey Knots!" amidst his words. There is also a note about who takes care of this cache. A family refills the water jugs and restocks the cooler with snacks provided by the local Safeway grocery. The snacks are expired or past their "Best By" date, but us hikers could care less. The fact they are not going to waste and instead are being used to feed the hungry gives me great relief.

It has been one of our hottest days on trail yet, and the heat of the day was getting to me before reaching this oasis. The late afternoon heat is sticky, and this is a great place to take a long break while enjoying the company, complimentary food, and clean water. Leah joins us after a short while, and we chat the afternoon away, talking of life and trail, trail and nonsense. I last saw them at the Toaster House, and this is a great place to enjoy a little reunion.

I sit at the water cache for three hours while Eighty-Five and Leah stay longer still. It is nearly six in the evening by the time I leave, but there are still a couple hours of light remaining before the sunset showing. Shadows of bushes grow longer as the sun moves toward the western horizon. Cows litter the land looking for food and water of their own, traveling in packs, walking the same path as me. Are the cows walking the hiker trail, or are the hikers walking the cow trail? I imagine trail crews putting up marker posts along an already existing cow path and declaring it to be the CDT out of convenience. The trail already follows plenty of vehicle paths, why not lead us down cow paths too?

I climb a small mesa, which offers a good view of yet another incredible sunset, as I chase the light before it disappears into the night. The trail twists along the mesa, and my eyes take turns glancing from trail to sunset, all the while looking around for a place to camp. Trail, sunset, campsite. Sunset, trail, campsite. Campsite, sunset, trail. Campsite. I find a flat, cleared out spot, and sit down to watch the sun set.

A patch of clouds illuminates in front of me as colors change by the second. I dare not remove my eyes from it in case I miss something spectacular. The phone and its camera are ignored. There is no technology in the world capable of capturing this moment nor my feeling of elation and purpose in this world. I am meant to be here, and Nature is thanking me for my attention. Sunsets in the desert are unlike any other, the flatness of the land and heat of the day give the sky a particular sheen.

I used to prefer sunsets over sunrises, but now I choose not to choose. Each holds its own bounty: the sunrise a reward for having risen early enough to witness it and the sunset a celebration of having been outside all day under the sun, waiting patiently for the moment that marks the end of another day well spent.

What I Learned in New Mexico:

1. *Stop and pet the cactus. Seriously though, stop. Otherwise, you'll poke yourself.*
2. *How to snap left-handed.*
3. *Don't squat with your spurs on.*
4. *How to track hikers from their footprints.*

V.

HIKER OR HOMELESS?

May 25 - Day 26

I intended to stop at this water source for the night, but I think I'll be changing my plans. It is early evening, and the cows have claimed this site as their own. No less than thirty of them surround the windmill and water tank. Water is spilling over the sides as a low sun reflects in its ripples. I approach the tank slowly, talking to the cows as I do, which appears to grant them no further comfort. Their groaning gets louder the closer I get, and some calves scuttle away with frightened faces. A bull stands ten yards away and refuses to move.

All of these cows look to him for their strength, and here he is demonstrating it by kicking his right rear leg and standing his ground. We maintain eye contact as I attempt to speak calming words to him, but I don't think he understands any of the noises coming from my mouth. His brow furrows, and he is threatening to charge, but I refuse to call the bluff. That's a big ass animal, and I don't want those horns anywhere me. I plead with the cows to let me through, but the bull remains stout. Desperate for shade, I retreat behind a small bush to eat a ramen before deciding what to do.

I weigh my options. Either camp here and hope the cows move along eventually, or take advantage of the last hour of sunlight and find a decent spot to sleep alongside the highway into Cuba. The map app also says there is a fairground two and a half miles away, claiming they have "tent camping to open soon." I am not one for watching cows lay on the ground, so I decide to carry on, accepting the risk of losing a decent campsite here in the bush. One liter of water will be enough to get me to the fairground, or even into town tomorrow if need be. The closer I get to town tonight, the less I will have to do into town come morning.

I turn onto the highway and begin the roadwalk, keeping my eyes peeled for a nice stealth camping spot tucked away on the side of the road. The sun ducks behind a mesa to my left, leaving me in the cool shade for the rest of the evening. I walk facing traffic and pump my arm to tractor-trailers as they pass. A couple of the drivers return my gesture with a pull of the horn as their massive trucks break through the wind. Soon, an approaching car slows to a stop beside me.

He is going the opposite way but nevertheless asks if I need a ride or anything else. I ask for some water, but he doesn't have any. We begin chatting, and it turns out he is a pastor who recently moved to town. I decide not to ask him why he got relocated. He continues to tell me where he came from as I eye the box of strawberries sitting in the passenger seat. The road has become oddly empty, what was a busy road twenty minutes prior, is now still and void of traffic.

I can't hold my breath any longer and finally blurt out, "Can I have a strawberry, please?"

"Huh? Oh yeah, of course. Take all of them if you'd like."

He offers me the plastic box. I take out the remaining five strawberries and hold them in my hand. I am usually desperate for food of any kind but always extremely desperate for fresh fruit. The first strawberry gushes as I sink my teeth into it, extruding juices onto my beard. This snack tastes better than anything I have eaten out of my food bag over the past five days. My decision to keep hiking is instantly rewarded by the appearance of this man and his leftover strawberries. As cars begin to approach, I say thanks once again and continue down the road toward town. My spirits are even higher, and I am glad to have made the decision to keep hiking this evening.

The CDT's opportunities for trail magic are scarce, but this is what makes these random acts of kindness so much more appreciated. Besides a couple of chance encounters such as this one, the only coordinated trail magic effort so far, was provided by a trail angel named Watermelon who supplied shade, snacks, and drinks about thirty miles south of Lordsburg. While I am thankful for the CDT's unfamiliarity, a lack of trail magic is merely a consequence of this, allowing the hike to be more authentic and challenging. Trail magic is never expected, and managing my expectations allows me to be much more grateful when I do receive a random offering from a stranger passing by on the highway.

I make my way toward the fairground, which is along the next road over to my right. I cross the space between the roads with much difficulty after it drops down into a small canyon before rising back up a steep dirt face on the far side. I climb over a couple of gates and duck through the private property as quickly as possible. This other road is a four-lane highway, and cars are moving much faster than the road I just left.

The sun is getting lower in the sky still, and I begin to wish I had stayed along the previous road. The off-trail travel to get to this highway took longer than expected, and night is approaching quickly. I eyeball spots off the shoulder of the highway, but more private property lines the road. Every ditch looks less appealing, and the shoulder is lined with trash. Cars speeding by would disrupt my sleep as well.

As the sun continues to drop further behind the mesa, I quicken my pace. The elation of receiving road magic fades, and my evening takes a sharp turn toward desperation as I look at the fences on either side of the highway claiming private property. "No Trespassing" signs hold threats that I am sure those hiding behind them are willing to enforce. If they see me. It is a curious feeling to be living in the wilderness, spending most of your time on public lands and then reaching towns that contain citizens who hold a sense of personal entitlement, declaring ownership of the land. This is not your land.

There being no way of walking to the fairground directly, my destination is still a mile or two away. I will have to walk around on the roads and won't get there until after dark. A mile outside of town, I see a man walking toward me on the side of the road. He has no backpack and does not appear to be walking straight. You can't be a backpacker if you don't have a backpack, and he should be walking on the other side of the road anyways. This is definitely not a hiker. He wears jeans and a t-shirt. A screwy smile spreads across his face as he stops to talk to me.

"Howdy," I say, greeting my highway-walking acquaintance.

"Hey, man. You see the sunset?" The words slur from his mouth as his eyes droop with drunkenness.

"Yeah. Beautiful isn't it?"

"Yeah, look at it. You know what it says?"

"Huh?"

"You know what it says? The sunset?"

I don't want to take my eyes off this guy and turn my back on him but glance over my shoulder quickly before replying, "It says the day is over."

"Nah, man. It says, '*be cool.*' "

"Sure. Different interpretations I guess," I conclude, beginning to walk away.

"Hey!" he calls, digging one hand into his pocket and reaching out the other, "I'm Dennis."

"Nice to meet you, Dennis. I'm Brian."

"Here man, take this." He pulls out two crumpled dollar bills and stuffs them into my hand.

"No, I don't need this," *but I could use it to do laundry tomorrow,* I think.

"Take it, man, take it. I want you to have it."

"I don't need this, take your money back," I say, holding out the crumpled bills.

"Nah, man, it's yours."

"Why do you want me to have it?"

"You're just like me, man. You're homeless."

"I'm not homeless, I have a backpack. And I'm walking to Canada," I add, as if having a destination makes me less homeless. I speak these words without much conviction in my voice. I mean, I am kind of homeless. Having a backpack does not make me not homeless, it just means I am more prepared to be homeless than some other homeless. Having a destination gives me purpose. I see myself less like a vagrant, sitting in one town begging for coin and more as a tramp, traveling up the country by foot to have an experience. Maybe I am homeless, but I sure am better off than this guy.

"What do you have in there?" he asks, a note a curiosity and jealousy in his voice.

"Tent, quilt, sleeping pad, change of clothes. I'm hiking the Continental Divide Trail," I say, hoping to strengthen my case of not being homeless. I might as well have said I'm walking to the moon.

"The what?" His face twists with confusion.

"The Continental Divide Trail, it's a long-distance hiking trail from Mexico to Canada."

"What? Why are you doing that?" he wonders.

"Because I like walking. Take this back, I don't want it. You need it more than I do," I implore, holding out the two bills. He takes them reluctantly and shoves them back into his pocket. I know he felt good about offering me money, glad to have something to give and hoping the karma would pay him double in the future. I nearly accepted the gesture out of politeness, and so he would not use those two dollars to buy a Steel Reserve tall boy come tomorrow afternoon. "Nice meeting ya, Dennis. Take care." I wave goodbye and turn to walk down the road, ready to get away from this character. It is getting late, and we both need to find a decent place to sleep for the night.

"Hey!" he yells after me.

I ignore him.

"Hey!"

Nope.

"HEY!"

I turn around. No way I can ignore that last shout.

"God loves you, man."

"Thanks, Dennis. He loves you too. Stay safe out there."

I'm not sure where he is headed to, probably to find a nice roadside ditch to sleep in, which I have no judgment toward having done that a few times myself.

I continue down the road and think about my homeless status. I made it clear that I am not homeless, but I cared more about demonstrating to Dennis there is a difference between him and me. My passion for the outdoors and my drive to accomplish a goal makes us different people. The backpack I carry contains all the things I need as a hiker, and after I get to Canada, I will find a place to live for the winter with proper walls and a roof.

Technically, I am homeless while I hike. My home is where I make it, and I close my eyes to a new place every night. A long-term traveler does not have a set home: the world is their home. Home may be a hostel for a night, a motel room for a couple of days, or a soft spot on the sand. I live on public land and underneath trees. I walk pretty much every day and set up camp when I'm tired. While my tent is not a house, it still checks all the boxes. It provides protection, comfort, and a space that is my own. If home is where the heart is, then I am always at home when hiking.

The ability to live without set accommodation is reassuring and oddly comforting. I know I can live out of a backpack for six months and be completely happy about it. Living with less gives me more opportunity to be in tune with my surroundings and more present. So many people live lives of unnecessary clutter and are numb to the world around them. They lose their grip on life while being suffocated under the weight of accumulated belongings, unable to live in the present, and constantly distracted by what they have or what they desire to acquire next.

Being homeless and constantly on the move gives me a perspective of what it is like to live on-the-go. You learn what you can do without, and you learn that you do not need much. While wearing this backpack, I have never wished to carry more. I always complain my pack is too heavy and am constantly thinking of ways I can lessen my load. All the things that clutter our lives hold us down and begin to strangle us after a while. The things we own contain us to one place or else instate a habit of accumulating more for accumulation's sake.

What *does* the sunset say? Is it saying goodbye after a long day of hard work? Maybe the sunset is a farewell for us to hold on to while it lies dormant before returning on the opposite side come morning. I rather the sunset not say anything, merely there as it is every day, not asking for attention or claiming itself worthy of our gaze. Perhaps the sunset is just marking the end of the day and approach of the night, a transitory show in the sky for all to see. The sunset speaks differently to all of us, an ever-changing tapestry that reflects the day's occurrences. Reinforcement that moments only last for the shortest while, and if you take your eyes off it, you are bound to miss something. Make sure you look long enough to appreciate what is there before it's gone.

I reach an intersection, turn off the highway, and begin down a county road. Half a mile earlier, I could see the grandstand of the fairgrounds but was incapable of getting there directly. This roundabout way takes me down a much quieter road than the previous highway. I get to the fairgrounds in the dark, and a locked gate greets me. An electronic marquee outside has a phone number blinking across it every thirty seconds. The lights dance in the darkness, advertising hours and events.

I consider climbing the fence but spot the cameras hanging outside the guardhouse. The keypad on the gate rejects my attempts at guessing the password. I have service, so I call the number. It rings all the way to voicemail. I hang up and call again. A disgruntled voice answers the phone.

"Hello?"

"Good evening, is this the number for the fairgrounds?"

"Yeah, whaddya want?"

"Sorry to bother you so late, but I'm a CDT hiker and was told we're allowed to camp at the fairgrounds. Is this true?"

"Yeah, that's right, but I live thirty minutes away so you'll have to wait for me to get there."

It is already past hiker midnight, and I wish to be asleep within thirty minutes.

"Can you just give me the code to the gate?"

"Nope, can't do that. And it's twenty dollars to camp."

"So, to be clear, if I want to camp here, I would have to wait thirty minutes for you to open the gate and pay twenty dollars?"

"That's right."

"Forget it. Sorry to bother you. Have a good night."

Click.

Worth a shot. Now I'm wishing I had camped with the cows. But I'm not with the cows, I am here. I look around and again consider climbing the fence. The door to the guardhouse is also locked. I would camp here, but the security light above me is really bright. I walk away from the entrance and back to the asphalt trail.

A couple hundred yards from the gate, I turn off the road and duck a few branches. There is another fence ten yards back, which I walk alongside for a minute before deciding on a sleeping spot tucked between two trees. I am exhausted and have walked plenty today. There are discarded beer cans amongst the grass and some other bits of trash. I unroll my tent to use as a groundsheet and inflate my sleeping pad.

Cars roll by in the dark. It will be very difficult for someone to see me in the night, behind these trees. After that conversation with Dennis about how homeless I am not, this exhibit may weaken my case. No matter. At least I will be warm thanks to the supplies I carry around with me in my trusty gear sack. I pull my buff over my eyes and let the occasional passing car lull me into a restful sleep.

What *is* the difference between a hiker and a homeless person?

Part Two: Colorado

VI.

COLORFUL COLORADO

June 3 - Day 35

I arrived at Trail Lake yesterday evening around six o'clock. The collection of lakes around me are half frozen. Patches of snow are scattered across the mountains. The trail leading here was nearly nonexistent as I followed the ridgeline for reference, making it twenty miles for the day. My legs feel strong, but a discomfort has been growing in my right shin over the past couple of days.

After making camp, I took an empty Ziploc bag, filled it with snow, and strapped it to my shin using a bandana. I am grateful for the snow despite post-holing across softening snowfields throughout the past few days. As the sun warms the snow, the top layer weakens. With every step across, I sink knee-deep into a wet icebox. This process is slow, tedious, and tiring.

I rise before seven this morning. Last night brought a short stint of rain, but when I peek out of my shelter upon awakening, I see blue skies to the west. I hear a rumble of thunder, look behind my tent, and immediately begin packing up camp. I pull the plug on a half-flat sleeping pad and look at the ominous clouds to the east. The sun is blocked from view, and clouds are moving at a fair speed. I figure it a good idea to get down from this ridge before the storm gets too close.

I hurriedly eat breakfast, stash a handful of snacks in my hip belt pocket, and get moving as quickly as possible. As I hobble away from the lakes, I feel tightness through my right leg culminating around the middle of my shin. I keep walking, making sure to extend my ankle as much as possible with every step. The only way it's going to feel better is if I keep moving in order to stretch it out.

It is still early, and the sun has not been able to shine through the clouds yet. Without a phone, I would not know the difference between 6:00 a.m. and 6:00 p.m., the dawn holding as many mysteries as the dusk. The trail leads slowly down the backside of the ridge as I walk from snow patch to snow patch. They remain scattered and are not deep enough to cause much trouble.

All too soon, precipitation begins to fall. It's not snow, and it's not sleet, but it definitely is not rain. Every so often, I check over my shoulder at the ever-approaching clouds, which encourage me to walk faster than I already am. Thunder is distinctly heard in the distance behind me, and I'm sure lightning is right there with it. As the tallest thing around on an open ridgeline, I need to descend this mountain quickly and safely. Ground between the snow patches is wet from melting snow. I walk swiftly but carefully as to not slip on the mud, snow, or rocks. Following an actual trail is obsolete now, I only care about losing elevation as efficiently as possible.

I check my phone to see the elevation profile and am not too pleased to see that the red line follows the ridgeline, not dropping below 11,300 feet for the whole day. I keep walking while considering my options, one of which is to set up my tent and wait it out. Another is to keep walking. I snack and continue moving, trying to ignore the thunder groaning louder still. Looking around at my surroundings, I notice ponds of stagnant water sitting along the trail. Trees are still a couple of feet shorter than me. The pain in my shin subsides, but I think the adrenaline is taking over. Adrenaline is telling me pain is not a priority. Common sense tells me to keep moving.

As so often happens in the mornings, I feel a poop coming on. I do not want to stop hiking due to lack of cover and know that I am just ahead of the storm. If I stop for a break, I will be overcome by it. This is a great plan to follow until it isn't. With every step, shit works its way further down the poop chute, and I need to stop soon or else risk making a mess in my only pair of shorts.

There is a stream running along the trail. I pause to fill a water bottle, but my body thinks it's time to unload what it has been carrying for too long. I grab my trowel and dash away to a patch of trees. I get as far away from the stream as possible but am quickly overcome by my bodily functions. The rain is getting heavier, so I keep to the trees for cover and protection. As I dig a cat hole, a clash of thunder rings so loud in my ears that I instinctively yell out and pull down my pants. Breathing heavily, I squat as a solid shit evacuates itself from my body, sliding out as smoothly as a penguin on ice.

"Shit, shit, shit," I say to myself while grabbing a handful of snow to wipe with.

I cover the cat hole with soil and snow before returning to my backpack. The storm has reached me. Out-walking it is no longer an option. I am under the clouds and in the storm. Rain collides with the ground and pelts my person. Thunder rings so loudly that I automatically curse with every reverberation shaking my senses. Although I have dropped five hundred feet since this morning, the trail is still following an open ridgeline.

I look up into a sky of gray demise. I have heard of storms being a problem in Colorado, but monsoon season is still a couple of months away. Nevertheless, these average storms prove to be much more of an obstacle when standing a thousand feet beneath the source. I take out my umbrella and pop it open. Before beginning again, I check the map to see what lies ahead. This ridgeline scramble must not be taken while a storm lingers above.

The map shows a bad weather alternate. A side trail dips down into the valley to avoid the high, open ridgeline lying ahead for the next twenty miles. I breathe a sigh of relief and see I am only just past the junction for this alternate. My phone screen is spattered with raindrops, but I still manage to identify the terrain that will lead me to this lower and safer route. I take to the trees for protection and move from grove to grove while avoiding the open grasslands full of standing water. Thunder and lightning continue to berate the sky around me. Speaking aloud to myself, I foster words of encouragement.

"There is nowhere to go but forward. I can either sit here shivering, under a tree or continue pursuing warmth, and safety," I say, repeating this mantra many times.

This personal pep-talk inspires me to keep moving out of fear for my well-being. Each strike of lightning sounds closer still. I jump at each burst of thunder as my heartbeat quickens. Rain pounds on my umbrella. I slip on a rock and fall to the ground. My foot placement has become erratic, and

I left the trail long ago. I see the valley I need to descend into and work my way to the edge of the tree. Sitting here to take a rest, I regain my breath, put my umbrella away, and wait for the next rumble of thunder.

Lightning strikes. I begin to count the seconds before the thunder follows.

One missi-*BOOM!*

I take off and run across the open field. My head remains down, and my already soaked feet sink into the wet ground. My shoes completely submerge, and I nearly lose one in the mud. Halfway across, another deafening *CRACK* fills the sky. I curse under my breath with every step.

"Fuck. Fuck. Fuck."

I manage to not lose my shoes and make it to the next grove of trees safely, out of breath, with feet only slighter wetter than they were a minute prior.

The map claims there to be a trail around here somewhere. I'll find it eventually. I stumble through the trees, navigating over patches of snow and fallen logs. The logs are slippery, and in my hurry, I continue to lose my footing.

"Slow down, Knots. This is no time to get hurt."

The trees provide cover, but I still want to descend into the valley quickly. The farther away from the ridge I get, the better I feel. The trail eludes me, but I know I am headed in the right direction. This valley and its corresponding stream will take me where I need to go.

A little further down, I stop to collect myself and hear voices in the distance. They are definitely ahead of me, but they also seem to be coming from across the water. Peeking through the trees, I spot a few horses on the opposite hill. There is a slight trail cutting along the steep terrain. If horses are on it, there must be some sort of path. I leave my not-a-trail and find a safe route down to the stream.

The water is loud and heavy from the rainfall, but I still hear voices. Down by the stream, I see two people on the far side, pulling their socks and shoes back on. They talk loudly to each other over the rushing water, unable to hear me approaching as I step into the stream.

"Hello!" I yell, making my presence known.

They turn around, looking surprised as I trudge through the water without bothering to remove my footwear. The water in the stream is much colder than the standing water I just walked through, but it's refreshing and cleansing. I think it impossible to get any wetter than I already am, which is partially true. It is, however, possible to get colder than I already am. I make my way up to the trail behind the couple.

Jack and Brenda are headed back to their vehicle after an overnight trip to Blue Lake, which is right where I turned off the high route and onto this alternate. They knew there was a chance of rain today and got up early to avoid the worst of it at higher elevation. I go ahead of them but talk over my shoulder as I walk in an attempt to distract myself from the chill spreading through my extremities.

Having not seen anyone in a couple of days, I carry the conversation. Jack is very interested in my hike and asks the same questions most people do: "What do you do for food?" and, "How many miles do you walk in a day?" I try to engage Brenda too, but she is behind him and having as much difficulty hearing me as she is in straining to keep up. I hear them both panting, clearly intensifying their walk to keep pace as I stride along, yelling behind me all the while, not losing a step, and still full of breath. I wonder how much longer they're going to hike this speed with me. I talk to distract myself and allow them to focus on their breathing rather than wasting their air yelling at me over the rain and thunder. For a short while, the hiking is secondary to the words I speak.

Eventually, Brenda calls for a rest, and I bid them happy trails, figuring I'll see them further along the path somewhere. I get out my umbrella again to protect my face and some of my backpack. It is nice to have a bubble of dryness within the umbrella's shield.

A little further down, I catch up to the guys on horseback I saw from across the river further back. There are three men and four horses: one from Ireland, one from Switzerland, and the other from Hungary. I don't ask where the horses are from. They have stopped at a stream crossing to let their steeds drink water. The men remove cigarettes from dry pockets beneath their ponchos. The one from Ireland has a small bag of loose tobacco and approaches me.

"Mind if I duck under your umbrella to roll one of these?" he asks.

"Not at all, as long as you roll me one too," I reply.

It seems appropriate to smoke with the cowboys. I'm not at all surprised they are subject to the habit. He slaps a pinch of tobacco onto the first paper and hastily rolls his own before taking slightly more time to roll mine. He is much more efficient at it than I would be. My fingers are so cold they would have difficulty functioning with such tiny commodities. I pull a lighter out of my hip belt pocket and light his rollie before bringing the flame to the one stuck between my lips. I take a pull and feel warmth spread from my mouth to my lungs.

They all met in Europe and came out here to do sections of the CDT on horseback. I learn that horses require at least fifty liters of water a day, something that is easy to accommodate for when the trail follows a stream and moisture is falling from the sky but far more difficult to subsist when trudging through the deserts of New Mexico. After visiting a few low cow troughs, they skipped up to the mountains of Colorado to ensure the equine's hydration as much as to keep their own spirits high in avoiding the heat and long waterless stretches.

I take off before their horses have their fill, so I don't have to worry about passing them further along. Also, this brief break has caused my blood flow to slow, and I begin to get cold again. I step through the stream and hurry down the trail, alternately blowing hot air into my hands as I go.

The rain does not let up, and I keep my umbrella out to ensure everything in my pack stays dry. After a short while, I feel my stomach growl. I have been hiking off adrenaline for most of the morning, and now that I am safely in the valley, my body is ready for some fuel. Not wanting to open my pack to the falling rain, I find a tree with a dense canopy. I dip off trail and dig into my food bag, pulling out some more snacks to shove into my hip belt pocket. I set off again quickly, afraid of getting too cold during the break.

The horses and those riding them are making a hell of a racket tromping through the woods and yelling to each other over the rumpus. They are half a switchback behind me when I step off the trail to allow them passage. Turning around to look at the horses, I see the Swiss guy pull hard on the reins to cut off this bit of trail, ignoring the switchback ten yards ahead. The horse obliges and stomps straight through the grass and trees. The other three horses behind him follow and, in doing so, leave a gaping hole in the trees, all in favor of cutting out twenty yards of trail.

I gaze after them with a look of incredulity but carry on behind their wake of destruction. The trail is a muddy mess from the rain and is getting destroyed from the unrelenting hooves of these heavy four-legged creatures. Debris lies scattered on the trail, and they posthole in the mud, leaving small round holes that drop six inches into the ground. Just before the stream, the Hungarian guy was walking his horse because it had already lost two horseshoes, and now he is atop driving it through the forest mercilessly. They leave an obvious trace in the woods, destroying undergrowth, and leaving discarded horseshoes behind.

I carry on behind the horses, slipping through mud along the way. After a mile or so, we reach a clearing, through which runs the Canejos River. The guys on horseback force their reluctant horses across it to the other side. While the horses trudge through in knee-deep water, the cowboys sit on top with feet much drier than my own. I remove my rain pants upon arriving at the river's edge. The water looks to be about knee high for me too. My feet are already soaked, what's a little more water going to do? The water is frigid but, at least the current is slow. My feet go numb with cold while my shin feels temporary relief.

Jack and Brenda arrive in the clearing shortly after me and once again remove their footwear before crossing. The horseback U.N. decides to head east toward the small, historic town of Platoro, to get rooms for the night and avoid the weather. It is eight miles away but a lot more manageable when you are not doing the walking. Jack and Brenda catch up to me on the other side of the river and inform me they are headed east from here as well, having parked their vehicle in town the day prior. I consider going into town, but the red line of the CDT lies to the west. I still have plenty of food and am not keen on bailing for the comfort of an overpriced hotel room just because of some soggy footwear.

"I'm gonna stay out here. I'll probably set up my tent soon to wait out the rest of the storm. It was nice to meet you two!" I say to Jack and Brenda, anxious to keep moving.

"Nice to meet you too! Do you mind if we take a picture with you?" he asks unabashedly.

"Uh, I guess not," I reply, slightly taken aback.

Jack pulls out his phone and begins toward me. Brenda doesn't move.

"C'mon!" he urges her excitedly.

She reluctantly slides over to my left, and he steps to my right. We put our arms around each other as Jack holds the phone in front of our faces. For a few moments, we are the best of friends. Partners of adventure caught in the same storm. Our paths crossed for barely an hour, yet the conversation we shared is likely to be stuck in Jack's head for the rest of the summer. This is his first camping trip of the season and quite possibly his last of the year. As a doctor, he finds time in the woods difficult to come by, and here I am, spending six months living out of a backpack. My walk amazes him and brings up torment in his mind.

The life of a thru-hiker is no doubt a romantic vision for anyone who thinks themselves stuck in the cog of society. The woods call to a specific nerve in a man's brain. The mind feels at ease in the woods, away from the noise of more settled people. The buzz of society fades to be replaced by the stillness of Nature. The wilderness fills a need for adventure and fresh air, imparting a return to where we began and where nothing matters. The "what if's" of life haunt a man, and the unattainable is desired, only because it is perceived at such. When the unattainable is seen in person, impossible is disproved, inviting a reexamination of the word.

Before they leave for good, Jack asks for my Instagram handle and if I need any more food. Wanting to contribute to my hike and well-being he offers up the last of their rations.

"We've got cashews, a few bars, trail mix… How about some dried apricots?" he asks.

"Ooh, yes please," I answer eagerly. It's barely noon. If I do stop walking for the day, it would be good to have some extra food on hand to manage my munchies.

"Babe, they're in your pack," he says, turning to Brenda.

His girlfriend once again obliges to his request and digs in her pack to retrieve the bag of dried apricots. She rescues them from the deep confines of her bag and drops them into my outstretched hands, clearly not wanting to contaminate herself with any of my germs or stench. I thank them for their kindness and conversation before turning west, alone once more.

This clearing is the low point of the alternate at 10,200 feet. The remainder of this alternate will take me up the north fork of the Canejos River, climbing 1,600 feet along the way. I am definitely out of the clouds, but they still enclose the valley and surrounding landscape. The rain is slowing down, but I should probably stop hiking. The trail ahead returns to the tight valley of the north fork to gradually regain the elevation I just lost.

My feet are numb with cold as I look around for a decent place to set up my tent, wanting to tuck away in my quilt to warm up. The ground all through the valley is wet, and there are very few trees large enough to provide protection from the rain. I hear a small stream rolling and walk toward it. There is an open spot nearby with pine needles covering the grass. If I set up my tent with the intent of waiting the storm out, it is a good idea to be near a water source.

The rain has stopped, so I take this opportunity to pitch my tent in record time. Having done this many nights over the past month, I am pretty adept at doing it quickly and efficiently. Unroll the tent from its stuff sack, lay it out on the ground, pull out the tent poles and snap them together, stake out the four corners before throwing the rainfly over top and attaching it to the stakes, stretch the fabric out and insert the last two stakes into the ground, making the rainfly as taut as possible. In less than five minutes, I have shelter. I throw my pack under the protection of the rainfly before dipping into the tent myself.

I leave my feet outside the tent and under the rainfly, removing my shoes and socks before turning my whole body inside the tent to zip it shut. I take off my wet clothes, put on a dry set, and slip into my quilt. The feathers warm me up, yet I still shiver. Fourteen miles in six hours, all done by one o'clock in the afternoon. It has been a wonderful day thus far, and I am thankful to at least be off the ridge. The rain starts again, and I listen comfortably to the patter of raindrops while removing stuff from my pack. Setting myself up for an indeterminable amount of time in this tent, I take out my food bag, journal, pipe, and weed.

VII.

QUIET ON THE CUT-OFF

June 13 - Day 45

I am back on trail after a week of zeros (days off) in Durango, Colorado. My best friend, Lisa, and her boyfriend took me in for the week and were incredible hosts. My shin splints were getting worse, and I needed to replace my sleeping pad and shoes. I also sent home my microspikes and water filter, neither of which I have any use for at the moment. The snow is melting fast, and the water in Colorado is so pristine that I see no reason to treat it.

I spent most of my time eating and relaxing, with bags of frozen peas strapped to my shins. The sky in town was smoky, and I kept an eye on the fires burning to the west. It was a dry winter throughout most of Colorado. While this favorably leads to less snow at higher elevations, it holds its downsides of not sufficiently wetting the trees and ground enough to prevent summer forest fires.

Fire season is in full swing, and while in town, a blaze in the San Juan National Forest continued to spread. Planes flew back and forth from the Durango airport to the flames, dropping retardant to protect homes and maintain its spread. I regularly checked trail closures brought on by this fire. The day before I returned to trail, the Forest Service announced more closures in the National Forest, including portions of the CDT.

If I wait any longer, I risk being forced to skip trail as the blaze spreads through more of the wilderness. We rise early the next day, and Lisa drives me back to Wolf Creek Pass so I can continue hiking. To avoid the fire closures, I will be taking the Creede Cut-off alternate trail, skipping the rest of the San Juans to reconnect with the CDT further north.

I am slightly disappointed to be missing out on the northern part of the San Juan National Forest. The southern San Juans were spectacular and thrilling. Ridgewalks provided endless views in every direction, snowfields stretched hundreds of yards across, and afternoon thunderstorms were a daily threat. The feeling of being on top of the world was enchanting. Running along ridges to avoid storms ensured my days were anything but dull. The San Juans remain a highlight of my hike, even if my time there was cut short.

Haze fills the sky as I walk the ridge past Wolf Creek Pass. The sun hides behind the smoke and glows orange. After a week in town, I am grateful to be back on trail once more. Refreshed and ready to hike, I take in every view and relish every step. My shins are still a little tight. I thought they would feel better after taking seven days off, spending those days icing and stretching for hours on end.

Shin splints can be a result of walking too fast, so I resolve to walk slower, at a 2 mph pace. This is a slower pace than what I normally walk, and I have to consciously pay attention to my stride. This controlled rhythm is more meditative as I focus on each step. My mind clears effortlessly after a week full of distractions and noise. Even without a watch or mile markers, I know a 2 mph pace compared to a 2.5-3 mph pace. Every hour, I check my progress and am surprised at how consistently I keep this rate of movement.

My first dinner back on trail is a ramen bomb (ramen and instant potatoes) wrap, complete with a dollop of avocado and topped with sweet and sour sauce. Even with the fresh avocado and complementary sauce, it is difficult to stomach. A week of eating non-trail food, and I have been spoiled to the taste and variety of other foods. A week of using a full kitchen, and I am reminded how disgusting cold potato flakes can be. Their blandness and dryness cause me to gag as I force the food down my throat, concentrating on the view more than the flavor, or lack thereof, of dinner.

The Creede Cut-off turns right, and I descend into the valley, waving goodbye to the tall ridgewalks of the San Juans. Having sent my microspikes home, I will not have to concern myself with any more tedious or sketchy snow crossings. This was another part of my reasoning to take a week off in Durango. Snow levels were already lower than normal, but I was willing to give the higher elevations more time to melt out so I could continue north without having to carry any snow gear. Unfortunately, the window between waiting for the snow to melt and avoiding wildfires is small, and one I was unable to time to perfection.

This is my first time being affected by a fire on trail, and consequently, my first time having to adjust my plans in light of it. Wildfires are much more common out west than back east, and it was only a matter of time before the random flame of chance interrupted my travels. There were a few fire closures in New Mexico, but these started after I made it through the affected areas. If I hiked into the San Juans before they closed them, I would have been caught in the middle of a closure and forced to shuttle out. The San Juans will always be there, and I am sure to have another chance to hike them in the future. I'm excited about the opportunity to walk through Creede, to see a part of trail most don't, and to visit a small historic mining town in the middle of nowhere.

***

June 14 - Day 46

The Creede alternate follows the slope of the mountain into the valley. Evidence of past fires leave their mark on the landscape. Tall, charred, and naked trees line the trail. Those remaining are ghosts of their former living selves. Most of the trees here are either burnt to the ground or brown and lifeless from the beetle kill that has left a scar across the Coloradan wilderness. These trees provide no shade and leave me open to the blazing sun. I take a break at a stream crossing to drink water and splash my face.

Evidence of beetle kill is displayed through some of northern New Mexico and across most of Colorado. Mountain pine beetles bore into healthy trees and consume their innards. The beetles leave eggs underneath the bark that will hatch after winter to produce hungry offspring. Hopping from trunk to trunk, these pests make their way through the wilderness, leaving a brown wave of decay behind. Cold temperatures kill them off, but with mild winters and less than average snowfall in recent years, the mountain pine beetle population continues to grow, threatening forests across the continent.

What used to be a luscious landscape, covered with thick green pines, is now a noxious brown that has taken over to dominate most of the forest. These dry trees are more likely to catch fire, adding to the forest's desire to burn and start afresh. Dead trees line the trail, and many lay across the trail corridor. Without regular maintenance, this infrequently used path is full of blowdowns. It is difficult to have a consistent stride when it's constantly being interrupted by waist-high hurdles and obstacles reminiscent of Nickelodeon game shows. Some of these are easy to step past while others mask the trail behind a thicket of branches.

A week without hiking, and my legs are well-rested. I do not see another soul on trail all day. After spending so much time surrounded by people in the city of Durango, I appreciate the solitude. Out here with the trees, life slows down. I know where I am headed and find comfort in my direction. Every step takes me closer to my goal. At the end of each day, I feel accomplished having traveled a great distance using just my legs. Our bodies are powerful, and so are our minds: thru-hiking tests both to great degrees.

Since graduating from college and hiking the AT, I have spent four years living in what most call the "real world." I despise this term and believe some use it as an excuse for why their lives are filled with such pain and unhappiness. The real world is what you make of it and shaped by your own experiences. My time in the real world has been incredible, and I am amazed at what opportunities exist. The real world is not a place where you busy yourself for the sake of having something to do. It is not a term to describe the superfluities of a boring life that repeat themselves every seven days, making money just to spend it. This is not the real world.

Living out of a backpack and walking all day through the wilderness is real. Drinking out of streams that flow through the veins of Earth is real. Being outside all day and night is real. I have never felt more connected to Nature and have never felt so alive. *This* is the real world. Trail life is simple and fulfilling. The normal world becomes the other world. The other world is a population I wish to disentangle from. I desire to remove myself from a life that is filled with chaos and suffering.

The real world on trail is simple and kind. The only question I absolutely need to answer every day is, *How far do I want to hike today?* Each night, I look at the map, mileage, elevation, and frequency of water sources, but all the research I do beforehand is merely information I can relate to when it comes time to answer this question. When sitting by a stream, contemplating the clouds and life, I ask myself, *How much further do I want to go, and what will best set me up for completing another section in the woods?* The answer to this perennial question changes every day, and I relish in solving the evolving equation.

Indiscernible signs and fallen trees,
That's how I like my CDT.

VIII.

HIKE NAKED DAY

June 21 - Day 53

I go into the roadside store at Monarch Pass to use the toilet and check the hiker logbook to see if anyone passed me while I was in Salida. Inside, I see a hiker with a fully stocked backpack. It is easy to spot fellow CDT-ers. We are the dirtiest, smelliest people in the store and hold an aura of freedom and carelessness unmatched by most. Sun-faded attire, dirty fingernails, and ragged hair are also dead giveaways. He is leaning over the log book, signing his name in the register. These log books are scarce on the CDT but provide hikers a stamp in time to prove their passing through. We look for familiar names, checking how far ahead they are before leaving our own scratch for those further behind still.

"Happy Summer Solstice!" I say, reaching out for a fist bump, "I'm Knots."

"Hey man, I'm Not Yet. Happy Summer Solstice to you as well!"

"You hiking naked today?"

"Hell yeah!"

"Me too! See ya out there. Or not."

He heads out while I take advantage of the plumbing and free toilet paper before starting my hike for the day. Outside, I walk along the road for a quarter-mile and locate the trail. It winds up the roadside hill and turns a corner. I pause here, hidden from the road, and psych myself up to strip down. Shame is one thing I decided not to carry with me on this hike. I hardly ever see anyone out on trail anyways. Hiking in the buff may be uncomfortable at first, but I know a way to work through discomfort. Do it and get used to it.

Today is day fifty-three on trail. After living out of a backpack for this long, you develop a certain amount of negligence toward things you thought were worthy of strain, consideration, or fuss. Hiker carelessness is a product of our transcendence into another dimension of freedom. The freedom of living in the woods and being relieved of all stress allows a hiker to blossom into another being. Circumstances or events that would normally cause a person bother or strain in the other world go unnoticed after a few weeks on trail. You realize what is worthy of worry and what is not. The opportunity to hike naked on the first day of summer is as much of an excuse to get naked as it is to demonstrate you have the balls to do so.

It's not like I wear much clothing while hiking anyways. Other than shoes and socks, my daily wear consists of a long sleeve shirt and Nike running shorts that have their own liner: two articles of clothing separating decency from disgrace. I pack away my shirt and stuff my shorts into the easily accessible bottom pocket of my pack. Within the first mile, I enter the boundary of Monarch Mountain, a ski hill during the winter and an empty mountain in the summer. Lifts and signs stand tall without snow covering the ground as I walk by on a service road. No lifts are running, and their summer operations seem to be nonexistent.

After crossing out of the ski area boundary, I return to public land and take my time striding across open ridgelines. The sun is getting higher in the sky, and I retrieve my small container of sunscreen. I can feel the sun on parts of my body that are not used to getting so much exposure. With more surface area in need of protection today, I am bound to run out of sunblock at some point. I should have bought more in town. My legs are heavily tanned, but my torso, thighs, and butt cheeks are in danger of serious sunburn.

Catching some color is on my mind, but I am enjoying

myself far too much to worry about a little redness. There should be more days during the year in which it is acceptable to not wear any clothes on trail. The air feels incredible on my naked body. The freedom of hiking in the woods all day is a great feeling, but without clothes, I feel even freer. This is how I came into the world, and this is how the rest of the animals live.

My body feels at home in Nature. Worries about being seen are far from my mind. Why should I be worried? It is they who should feel shame for wearing clothes on Hike Naked Day. The summer solstice is the longest day of the year and should be celebrated properly. The smile on my face grows wider with every step. I have the whole ridgeline to myself this morning, and I stride along happily wearing only socks, shoes, sunnies, and backpack.

The trail turns right off the ridge and descends 2,000 feet into the valley. There is an alpine lake with a stream flowing out of it. As I approach the stream, I hear voices a little further down. I consider pulling on my shorts for only a second, quickly realizing it'd be a silly thing to do. These may be the only people I see all day, and I want someone to bear witness to the crazy naked hiker on trail. As the voices get louder, I see three clothed hikers with backpacks off a little ways down the stream.

I step lightly down trail and take a water bottle out of my pack while I walk. Before getting too close, I step off trail to fill my bottle from the stream flowing strongly out of the lake. The water in this state is pure, and when it comes directly out of the ground at such a high elevation, the risk of catching something is low. I have yet to get sick and will probably continue going unfiltered until I do. I take a few swigs directly from the bottle and splash my face with the alpine water.

Stashing away my water bottle, I bounce down trail toward the hikers who are not participating in Hike Naked Day. As I approach, I see them all force their eyes elsewhere rather than completely acknowledge my presence. By doing this, it is clear they have noticed my approach. I figure I should talk first and give them a hint as to why I'm hiking in the buff.

"Happy Summer Solstice!" I say when I get close enough.

The lady keeps her eyes diverted but says, "Oh, is that what's going on?"

"Yeah! It's Hike Naked Day!"

I must not be the only bare hiker they have seen today. We exchange a few more words, but they keep their heads turned away the whole time. I'm not sure which way they are going but wish them a great day, as I continue moving down trail. Once out of earshot, I laugh aloud to the trees. They dance in the breeze as my bush does the same. The excitement of the day has me ready for lunch, and it is about time for a break.

The trail will be taking me past another lake soon, and I figure it a great opportunity for a swim on this hot summer day.

I took a shower in town yesterday and feel too clean for my own good. Already void of clothes, I drop my backpack on the shore and remove my footwear before wading into the lake. Skinny dipping is preferred when hiking. Wet shorts are uncomfortable to walk in, and they take too long to dry. The water is cool and refreshing, a comfortable temperature when compared to the warm air.

Back on shore, I sit in the sun to dry out. The body has a curious way of evaporating all moisture or else taking it in to hydrate itself. Within minutes I am dry. Sitting on the grass, I take out my food bag and prepare lunch. Today the lunch special is a refried bean, Frito, and couscous burrito. Recently, I have gotten into the habit of packing out refried beans. They are heavy compared to other trail foods but packed with protein and quite delicious. In town, I use my small knife to open the can of beans and dump them into a double-bagged Ziploc. Some stores carry dehydrated refried beans, but I have not seen them since Silver City, New Mexico. My vegetarian ways have me hungry for protein, and this is one way of getting it.

Next, I take out a pop-tart and a jar of icing. Icing is a great topping on most things and provides a sugar rush that will consequently have me crashing an hour or so down trail. My final dish for lunch is a packet of cold oatmeal mixed with a bit of granola. I pour some water into the packet and scoop the contents into my mouth. Sometimes, I snack while walking but prefer to sit down. I am better able to appreciate every bite as opposed to doing two things at once and forgetting to chew. Too often, I find myself holding a wrapper, wondering where the granola bar went that was there a minute prior.

I relish my time by the lake. Hidden from view of the trail, I will not be seen unless someone walks down here the same way I did. Even if I do not see another CDT hiker celebrating the same cause, the freedom of being without clothes is a freedom I am thoroughly enjoying. Being alone makes certain things more challenging than they would be when surrounded by others. The support of other souls is absent. With only my thoughts and self to rely on, encouragement comes from within. Whether I am hiking naked on the first day of summer, leaving town after a day off, or walking in the dark to find a decent campsite, personal motivation carries me forward.

As a backpacker, one of my goals is to do with less: less gear, less stress, less things. After today, I know I can do with less clothing too. While clothes are certainly necessary when the temperature drops, they are not as necessary during the warmer months. Other hikers may look at me today with disgust and shock, but I now appreciate the protection of clothes even more: protection from bugs, the sun, and accusatory glances.

It is with a smile that I get back up and carry on down the trail. At least now, I am under the trees at lower elevation and don't have to apply sunscreen to save my pale skin from harmful rays. The shade provides a refreshing change of scenery, and the smile on my face remains. Walking every day can be monotonous, but at least today holds its own unique challenges. I continue north without clothing, feeling right at home in Nature.

ALL ABOARD THE COLORADO TRAIL

Good morning, and welcome to the Colorado Trail section of the Continental Divide Trail. For the next four hundred miles, these two wondrous trails become one to take you across the breathtaking mountains of Colorado, over majestic peaks, and through winding forest trail. Cruising altitude will be around 10,500 feet, with intermittent descents into the valley, along with frequent climbs up to, and well above, 13,000 feet. Portions of this trail are open to mountain bikes, so please keep an eye out for these vehicles and be ready to step off trail at a moment's notice.

If you need oxygen, no help will come from above. Have a rest to gain your breath before continuing. Afternoon storms are common at higher elevations, so try to avoid hiking on exposed ridges when dark clouds linger above. Descend below treeline if you hear thunder or see lightning. Familiarize yourself with emergency procedures, and always have a plan.

Please keep your hip belt properly fastened and smoking paraphernalia nearby. Remember, marijuana is legal* here in the state of Colorado, so blaze that shit up, cowboy! We are in the great outdoors, after all. The featured beverage for this trip is our very own, crisp, cold, clear, Colorado mountain spring water. Snowpack at higher elevations continues to steadily melt in the warm summer sun as streams gush with Mother Nature's finest.

When arriving upon water sources, be sure to look to your left and right. Established campsites are located in these areas and can be a great place to enjoy breakfast, second breakfast, brunch, lunch, water break, sit-down, snack, safety meeting, first dinner, second dinner, or dessert.

Properly abide by Leave No Trace Principles, and keep electronic devices in airplane mode at all times. Thank you for hiking with us today and happy trails!

**Smoking of marijuana in any national forest is illegal. Smoke at your own risk.*

IX.

PEAKS AND PASSES

June 30 - Day 62

The red line descends into the valley before climbing back up to Argentine Pass. I forgo the official trail in favor of a scenic alternate. The Argentine Spine alternate route remains high, running along the ridge, but the total change in elevation is still comparable to that of the lower route. There are not any water sources on the ridge, but I camel up at the last stream and conserve my supply well, considering the demanding terrain of the alternate. Each peak feels higher and higher, while the descents between seem to drop deeper with every passing saddle. The whole spine is void of any trail, not that I need one anyway.

I am atop Argentine Peak, standing at a handsome 13,743 feet. The Argentine Spine lies behind me. Looking to the south, I track my path along the highest ridgeline around for miles. From this distance, it doesn't look that bad. From this distance, the next five miles don't look that bad either, though I know it will be the most difficult part of my day.

The wind on Argentine Peak is brutal. The summit hosts a small wind barrier built out of rocks and a post bearing the first CDT marker I have seen all day. It is around noon, and people are clambering up to the top from all sides. This mountain is a thirteener and has a logbook tucked away in the rock wall. I don't stay up here long. It is crowded, and I'm getting cold. I sign the tiny logbook before carrying on toward the pass.

There is still no discernible trail, but I aim for the cars parked at the pass. The wind attempts to push me over and knock me sideways. I yell frustrations to the breeze that are blown away immediately. As I approach the pass, I begin to hope that a kind someone will have extra snacks. I will be in town tomorrow, but my food bag is uncomfortably low already.

I reach Argentine Pass, where the alternate rejoins the CDT. I stand behind one of the five Jeeps parked here, using the vehicle as a wind block. Alas, I am taller than a Wrangler and have to lean over on my trekking pole to shield myself from the unbroken onslaught of wind. The people here mill about the dirt parking area for a few minutes before tiring of the wind themselves, retreating to the protection of their vehicle to take in the scenery. I take out my food bag and attempt to look as beaten down as I possibly can. It comes pretty naturally at this point. I eat a crumbling Pop-Tart and catch my breath.

One girl notices me. She gets out of her Jeep and approaches. Protruding from her shorts is a glorious pair of legs, goosebumped from the gust. She wears a sweater and look of concern. Blonde hair whips around her face as her blazing blue eyes meet my own. I have been sent an angel.

"Do you need anything?" she asks concernedly.

"Snacks, if you have any? Maybe some water?" *A kiss? I'll take whatever you want to give me.*

She opens the back of her Wrangler and unlatches a cooler, revealing a lone beer and a container of hummus. She also pulls out a bag full of bars and other snacks, allowing me my pick of the spread. I am so relieved and thank her repeatedly. My desperation and hunger must have been obvious. The kindness of people is great, yet even greater when they offer help. I explain to her my food supply is running low and will not be able to get to town until tomorrow.

"Are you doing Grays today?" she asks.

"Yeah, I follow this ridge all the way up there and will descend the other side. Have you done it before?"

"No, not yet. How long have you been hiking?"

"Today's day sixty-two," I casually reply.

Her mouth opens in amazement while her eyes do the same.

"Day sixty-two?? Where'd you come from?"

"Mexico."

Her mouth drops even further.

"How much longer are you hiking for?"

"Well, I'm on my way to Canada, so probably another three months."

Suddenly, everyone on top of the pass is out of their vehicles, and I am surrounded. They approach me with snacks of their own offering as they explode with words of disbelief and questions that have all been heard before.

"Are you following a trail?"

"How many miles have you hiked since Mexico?"

"Sixty-two days?"

"Is that all you're carrying?"

"Where do you sleep?"

"How many miles have you hiked today?"

"Do you carry a gun?"

"How about a knife?"

"You want some food?"

"How many bears have you seen?"

"Sixty-two days??"

I take a bar from one of the other angels and eat it while answering as many questions as I can. Each time I finish a response, others rush to ask one of the many questions bouncing around in their mind, hardly listening to whatever else is being said. Answering these common questions is easier when people feed me while I do it. I look around again. Half of the people are close to me, while the others stand nearby enough to hear everything being said, unable to think of any questions themselves, stupefied into disbelief.

"Hey, Knots! Did you take the shortcut?"

Another two hikers have joined the circle, and one of them is Captain Jackson, a flip-flop thru-hiker I met back in Creede. He began his hike in southern Colorado and will be flipping down to Pagosa Springs once reaching the northern terminus to continue south from there. Captain Jackson is out for a day hike with his brother, who lives nearby in Denver.

"You mean the spine?" I answer. "It was incredible! Hardly a shortcut. Probably more elevation change than the trail that went down into the valley. Which way did you go?"

"We went through the valley, and it was fine. The water sources were nice." He looks disappointed at having not walked the Argentine Spine, but he is also hiking with his brother for the day, so he was forced to take the easier, more accessible route of the two. For most, it can be difficult to hike with friends or family who want to come out for a few nights or even just a day. Your mileage has to be cut in half, otherwise they get overexerted, tired, or hurt and may never want to hike again. "Was there any water up there?"

"There was a patch of snowmelt at the beginning but nothing the rest of the way. I carried four liters, and it was plenty."

The group around me has gone silent at this casual exchange between two people with large backpacks on who have just happened to run into each other on this pass.

"Wait, are you hiking the CDT too?" one of them asks, looking at Captain Jackson with the same eyes they were gazing at me with, clearly amazed that I'm not the only crazy person hiking this trail, and there are others just as sane as I. Their attention is taken by the arrival of another hiker, and I jump at the opportunity to leave.

"Thank you again for the food," I say loudly, backing out of the circle, "you really saved me. I gotta get goin' before those dark clouds threaten the rest of my day."

They wish me luck as I raise two fingers over my shoulder in reply.

There is one more string of mountains to connect before I reach the saddle beneath Grays Peak. The wind continues to blow me sideways, but my spirits are high, knowing I have plenty of food and do not have to worry about rationing any of my snacks. I was cranky this morning because I was slightly underfed and a little dehydrated, but sometimes sacrifices have to be made, and the pay-off of walking the Argentine Spine was well worth it.

The next ridgeline is just as challenging as what I hiked this morning. The rocks that cover these mountains are large and stick out at odd angles. Every step must be taken carefully on the scree as the ground shifts under my feet. I look ahead to see what the easiest route is but also force my head down, cautious not to misstep. Storm clouds rest in the distance, far enough to ignore but close enough to keep me moving.

To my right, the mountain drops a thousand feet off the side of the ridge. The left is slightly more forgiving, but there are minimal traces of travel on the rocks. Every so often, scree slides down the slope as I do my best to walk softly on the slack talus. Surprisingly, I am not the only one on this route. A few small groups are navigating the same spine. We walk more carefully with others around, not wanting to send a rockslide each other's way. A pair of hikers twenty yards below pause as I walk past above them. When I am far enough along, they continue on.

"Thanks!" I shout over the sound of shifting rock. They wave hands back before moving forward.

A little further ahead, I spot another group of three crossing the saddle beneath Grays. This time I take a rest, allowing them to keep their momentum. When they get closer, I notice their age. They must be in their sixties and look as comfortable on this exposed slope as they do on a dirt trail under the trees.

"Did you make it to Grays?" I ask when they get closer.

"Nah, it didn't go. We turned around short of it. Worried about the weather and getting back to our car before dark," the man in front answers.

The clouds are beginning to look ominous, but I don't have any other option at this point.

"Bummer, I'm sorry to hear that."

"It's all good!" the other replies with a big grin on his face. "We've done it before, just can't move like we used to."

"Hey, at least you're out here! It's a beautiful day for a hike."

"Where did you come from?" the first man asks.

"Mexico."

"Huh?"

"Like fifteen miles back."

"No, I mean, what trailhead did you start at?"

"Um, no trailhead. I just camped on a ridge about fifteen miles back. Did the Argentine Spine, and now I'm here."

He wears a confused expression, but the man in front is still smiling.

"You must be doing the CDT!" he interjects excitedly.

"That's right!"

Even a casual fourteener bagger has no clue what the CDT is. I'm getting anxious to summit Grays, so I leave his friend to educate him on the particulars of the subject. I have had my fill of answering questions for the day and take off as the clueless one yells after me, "Where are you going?"

"North!" I call over my shoulder.

The grace at which I navigate the scree surprises even myself. Each step is placed carefully, as I eye the rock I am about to step on while also evaluating those that surround it. Most rocks move a smidgen when I step on them but prove to be sturdy enough. The path up the side of Grays Peak is faint, and I try to follow it whenever it is obvious. The switchbacks turning back and forth are tedious. Instead, I climb straight up the side using my hands to help pull myself over some of the taller rocks. The wind blows around me, and the sky is hidden behind dark clouds.

This is a steep climb, and it being my last one for the day, I am exhausted. The peak is not visible due to the steepness of the climb, but every step takes me that much closer. I talk to my mind and body, speaking words of encouragement for my ears to hear.

"These legs are strong. This body is strong. This mind is strong."

I repeat this mantra aloud and take a breather every two hundred steps. Soon, the climb begins to level out, and I see the top of Grays Peak ahead. I get to the top and let out a yell of joy at having reached the highest point on the whole Continental Divide Trail at 14,278 feet. There is another man at the top, taking in the view toward the south.

"Congratulations!" he says.

"Thanks!" I respond, breathing heavily. "Do you mind taking a picture of me?"

Whenever I come across someone on a peak or pass, I make sure to get a decent photo that is not self-timed. The view is spectacular, and the panorama atop this peak is well worth a whole day of ups and downs. From here, I can see for miles in every direction. The clouds sit around some peaks in the distance, but this one remains clear, for now. The guy takes my picture, and I head over to the rock half wall built to protect hikers on windier days. Three more people are sitting down here, each drinking a white-mountained Coors in celebration of the summit. It is their first fourteener as well. We congratulate each other, and I take a seat to join them.

What a day. I have been hiking since 5:15 this morning, and it is now 2:15 in the afternoon. Nine hours to go a difficult eighteen miles with only a couple of short breaks. Now I can enjoy the view a while and take the next four miles slowly. This is the first fourteener I have summited but could care less about the height of this beauty. Tall mountains are all around, and I trace the day's path with my eyes, following the twisting ridgeline back until it is lost in the distance. I smile and reflect while savoring the perch atop Grays Peak. Right beside Grays is Torreys Peak, another fourteener. It's a few miles away, off-trail, following a descent to the saddle before rising back up to 14,275 feet. I consider bagging it for a moment but decide my body has had its fair share of descents and climbs for one day. Besides, there are more dark clouds in that direction, and I am more than satisfied with the day's work.

X.

JUST KEEP WALKING

July 10 - Day 72

My parents dropped me off at Berthoud Pass a couple of days ago. The week vacation was a nice break from trail, and I am thankful for them having come out to see my brother and me during this adventure. Nick took a bus from Steamboat Springs to Denver, and we dropped him back off in town before returning to Berthoud Pass, where I got off trail. They slackpacked me for one day, during which I did twenty miles, we visited a few trail towns, and spent three days in Rocky Mountain National Park. It was also my first time seeing Nick since we stopped hiking together back in Grants, New Mexico. His FKT attempt ended up not working out due to fire closures in the San Juans, but he is making great progress and should be reaching Canada within two months.

I am always relieved to be back on trail after taking a few zeros, but I find my motivation lacking. From the beginning of the trail, I was hiking to Lisa in Durango. After that, I was hiking to the Interstate 70 crossing just outside of Denver to meet up with my parents for another brief break. Between these two checkpoints, I kept motivation by checking the World Cup schedule to see which games would be on during my town visits.

These events encouraged me to keep walking and to get somewhere by a certain time. The World Cup is over, and everyone I know is behind me (except for my brother, who is only getting further away and guaranteed to finish at least a month before me). My only remaining deadline is that of making it to the Canadian border before winter arrives. A land of unknowns is all that remains. Unknown territory, unknown people, unknown experiences. The future is full of possibilities.

I walk every day because I know I must. If I don't, I will only feel worse for not making any progress. There is nobody out here to talk to, just the stars shining in the dark or the air always encircling me. Walk and eat. Eat and walk. I can even do both of these things at the same time. The trusty hip belt pocket carries anything I may want within easy reach: phone (map, music, podcasts, time), snacks, chapstick, pipe, and weed. Everything else is on my back.

It has taken me a while to get through Colorado. It is a well-populated state and seems out of place when compared to the rest of the trail. The outdoorsy nature of Coloradans is impressive, and many recreational opportunities are taken advantage of by the locals and tourists alike. The strangle of this state has been holding me tight thanks to its frequent (and expensive) ski town resupplies, along with the twelve zero days I took to be with friends and family.

Though the enchanting views certainly do not get old, the days are beginning to blend together. I am walking another ridgeline, moving mindlessly as my legs carry me onward. The sun sits high in the sky as the hot, unforgiving July air tires me out. Above treeline and far from any shade, I hike most of the morning with my umbrella out. One good thing about the afternoon is it is often accompanied by dark storm clouds that block the sun and allow short stints of shade. Thankfully, these clouds never do more than threaten a storm or release a few rumbles of thunder.

At least at elevation, I can see for miles in every direction and can evaluate the possibility of a storm actually impeding my progress. I see storms approaching and notice every shift in the weather. On these ridgelines, I'm the tallest thing around for miles. Encroaching clouds, which rumble in the sky, push me onward and provide a little more motivation.

Thunder shakes the ground and my nerves at the same time. With every step, I am in awe of what surrounds me and how there are so few people out here to appreciate it, making my experience that much more enjoyable.

I remove my phone from my hip belt pocket and take a picture. Compared to the scene before me, this digital interpretation is a disgrace. Why do I bother taking photos? It's not like I spend my whole day, phone in hand, snapping thousands of pictures, but every other one I capture is disappointing in light of what stands before me. I take my phone out on occasion in an attempt to capture the awe-striking beauty of Nature, but the result fails to encapsulate everything I'm able to witness through my eyes.

A picture may be worth a thousand words, but some moments and experiences cannot be expressed with words. The smell of a thousand pine trees. The sound of rain droplets being released overhead. The whisper of wind past the branches of a fire-scorched tree. Much is lost when looking at this scenery from behind a screen. The natural world's enchantment is negated when viewed this way.

Lighting is not always perfect, and clouds that give the sky certain character are cursed for blocking the sun and ruining a photo. The desert is unique, down to the cracks in the dirt and the dust that rises with every step. Trees covering a mountainside are each individual strokes on the canvas of

Nature. Such things are missed in a picture. A crude snapshot of such a spectacular scene is hastily captured. Each mountain top view deserves a skilled artist with a delicate hand, keen eye, and many hours to attempt to brush Earth's beauty.

I have not been taking many photos in the first place, but my frustration of not being able to properly capture these stunning scenes influences my conscious decision to take my phone out less often. This resolution to stop taking so many pictures also comes from my desire to take in the present scenery as much as possible while I still can. Every day I spend on trail is a treat, and while I want photos to share and remember the hike by, I need to enjoy it while it lay in front of me.

The purpose of taking photos is to remember what I did, but it also holds secondary motivation in showing off to others where I am, what I'm doing, and what I'm seeing. My pictures end up as eye bait for those brainlessly scrolling by on a screen, searching for a combination of colors that will take them out of their present and into my past. I feel an obligation to share a bit of this hike with family, friends, and random strangers on the internet who are all supportive of this endeavor, but it forces confirmation of beauty to be postponed until I get around to posting a picture online. The views I observe are spectacular, but a part of me still craves validation from others.

There are some things I do not allude to with these words, and many experiences have been left out on purpose. Reading a book or scrolling through thousands of pictures will give you no comprehension of what a thru-hike is actually like. No video, film, or slideshow will effectively demonstrate every unique emotion felt during a six-month trek. These words I share are merely those of motivation to pursue a dream and encouragement to go outside to appreciate the wonder of our country before it's gone.

I am twenty-six years old, and though I do not have my life figured out, I prefer not to at this point. Youth is still on my side, and while out on the Continental Divide Trail, it is difficult to wish myself elsewhere. All around me, I have people jealous of my nerve and daring to enjoy such an inconsistent life, but I would live it no other way. At this age, I have no desire to propose a course for the rest of my life. I do not want anything to be set in stone, and I still look forward to the future because of its uncertainty. My whole life is ahead, and I refuse to set it in perpetual motion without knowledge of what encourages my motion, to begin with.

I am not completely happy with every bit of my life, and that's okay. I could make the argument that I shouldn't be happy with where I am, which is also fine. To be human does not mean to be happy. What have I accomplished in life to be proud of? An expensive piece of paper that has barely seen the light of day, a few thousand miles hiked, and a handful of quality friends? No, there has to be more to life than this. There has to be more to life than just finding someone you can get along with in order to begin reproducing. There has to be something more to accomplish before gaining true satisfaction. More to see, more to try, more to experience.

If I am not happy, I will concern myself with doing something that encourages joy. Complete happiness is rare, and while I believe it can be achieved in moments, it is far more difficult to consistently find and feel. This want will be my cause for growth and motivation to learn. Maybe I am dispirited when it comes to my other life, if only slightly, but I rather be aware, confused, and striving to improve than unconscious, ignorant, and content.

The trail has taught me many things, but one of the lessons I learned on my first thru-hike was the Backpacker Law of, "The Trail Provides." If you are running low on food, you will come across a day hiker offering up the remainder of their snacks. If you are looking for a campsite, one is likely to appear. If you are conflicted, you will find resolve. Whenever I start to feel lonely, I meet a local in bar or a fellow hiker on the trail. Days when I do not see anybody, I focus on myself and bask in the silence of Nature, appreciative of the solitude. Often, I do not know what I need until the trail brings it to my attention. The unpredictability of a hike is one of the things that ensures every hike is different. Off trail, we live lives of beautiful repetition. On trail, our lives are able to thrive under the pressure of pure chance.

Most of what happens on trail, a hiker has no control over: weather, terrain, the trail's condition, when one of those many passing cars will pull over to offer you a ride into town. Things will happen as they do, and as a hiker you must be flexible, always willing to adjust your itinerary. When in town, it's important to check miles and stock up on food. These are a couple of matters you can control as a hiker. It is necessary to remove yourself from the trail to analyze what lay ahead and plan appropriately. My life away from trail was lacking destination and these regular check-ups of progress. Hiking allows me an opportunity to view my life from a different perspective.

Going with the flow gets you somewhere, but without even a little bit of planning, the flow may take you into a current that you do not want to be in. Be careful of going with the flow too much. You have control of your life, make sure to nudge it in the right direction.

Part Three: Wyoming

XI.

HALFWAY(ISH)

July 18 - Day 80

I hoot and holler at having reached another border. My shouts split the silence and carry through the trees before nestling into the long grass. Two states down and 1,287 miles hiked to this point. Sitting at the Colorado/Wyoming state line, I pack a celebratory bowl and prepare breakfast.

The border is marked by a line of rocks, with more rocks on either side spelling out "WY" and "CO". There is a sign on the tree that says "Wyoming state border" and is accompanied by a couple of license plates. A Wyoming license plate hangs on the north side of the tree while a Colorado plate is nailed to the south side. I smoke a bowl and eat cold oatmeal while sitting in Colorado a while longer. The reflection on what I have accomplished thus far as important as preparing myself for what is to come.

Having only hiked two miles to get here, the sun is barely peeking through the trees that stand tall around me. There is no rush. Twenty-one miles to Battle Pass and the hitch into Encampment, Wyoming. I don't want to get there until tomorrow morning, so I will go another eighteen miles and camp a few miles before the pass.

State borders are great for morale. Though a mere line in the ground, it means so much more to a thru-hiker. With the trail only going through five states, each one is roughly twenty percent of the hike (with Montana and Idaho containing slightly more and less, respectively). I am excited to leave Colorado in my wake even though north of the Colorado Trail the CDT as I like it returned. The mountains were incredible, and the ridgewalks magical, but I was in this state for a month and a half. With a couple of trailcations, I certainly took my time to get through it and cherished every step, town, familiar face, and stranger I saw along the way.

Ski towns line the trail throughout Colorado. While expensive, their charm was undeniable, and they often reminded me of Mammoth Lakes. I found similarities in the towns and people alike. The green ski hill off in the distance looms over the whole town, visible from every street corner, dangling the inevitability of winter over everyone's head. While vacationers and locals alike enjoy the overwhelming amount of summer activities, ski runs void of trees are visible in the background forcing many to daydream about sweet, soft turns during the chill of winter. The impending end of summer allows you to appreciate its duration, knowing soon enough everything will be covered in a blanket of white. After going through Steamboat Springs, the last resupply in Colorado, I was ready to get away from the luxury of these towns and the crowds that came with them.

Wyoming is the least populated state in the country, and many gems wait to be discovered within its depths. With the Rockies behind me, I now look forward to the Great Divide Basin, Wind River Range, Yellowstone National Park, and all that lies between. This hike has taken me through many unknowns, and everything here north is an exploration of strange territory.

Eighty days in, and this is my life. I walked here from Mexico. I open the map app and load the Wyoming section. The red line squiggles 507 miles from my point on the border, up and out the northwest corner of the state. *Just 500 miles*, I tell myself. It should take no more than a month to accomplish this. Now that I have hit this mental halfway point, my confidence in reaching Canada increases.

In New Mexico and southern Colorado, whenever a passerby on trail or townie asked where I was going, I would often reply, "North!" With my goal becoming more achievable by the day, I now respond by saying, "Canada!" I do not want expectations to get in the way of this adventure, but if stating my intention clearly gives me more drive and encouragement to reach the border in time, I will continue to speak this truth in order to reach my goal.

I pack up my bag and shoulder it, ready to be done with Colorado. I plant a dry kiss on the border sign and step into Wyoming with a smile on my face. As soon as I cross the border, the trail thickens with overgrown grass, and blowdowns become more frequent. The trail is wet, and I walk through muddy grasslands, getting my feet thoroughly soaked in the process. The last few days have indicated a change in environments. The forest thins, allowing more sunlight to shine through gaps in the canopy. Mountains fade into hills, leaving my legs thankful. The Great Divide Basin is less than a hundred miles away, and the tall peaks of Colorado continue to drift further behind me.

Colorado had the most well-maintained trail so far. Outside of the Wind River Range and Yellowstone National Park, I expect Wyoming to be more rugged. I am mentally preparing for more roadwalks and fewer ridgewalks, smaller towns, and fewer people. There was some rain yesterday, but today looks to be magnificently clear. The blue sky shines behind an emerald canvas of pine. Not one cloud in the sky, and not a single person on trail beside I.

Going a whole day without seeing another soul is normal and has been happening fairly often recently. I spend my days with whatever is bouncing around in my head while also taking the appropriate time to clear my mind of all thought, practicing a walking meditation. Whenever a thought approaches, I notice it and let it go. I concentrate on my breathing. In through the nose, out through the mouth. In through the nose. Out through the mouth.

In.

Out.

In.

Out.

My mind calms with every breath. Thoughts lessen with every step. There are plenty of minutes in the day to reflect on the past or ponder the future, so I remind myself to take enough time to be present. Concentrating on my breath is the easiest way for me to accomplish this. I notice the smell of Nature and the very thing that keeps me alive. The simple act of breathing is something we do not have to think about, which is precisely why we should think about it. Allow time to indulge in breath.

Tall guardians lining the trail release oxygen as I take it in slowly and appreciatively. With every lung full, I am as grateful for the fresh, clean air as I am to be in a land void of human destruction. The only trace left by humans is the narrow path below my feet, along with the occasional wooden post stuck into the ground or marker nailed to a tree. Nature is most beautiful when she is left to be, untouched by the harmful hands of greedy men. Birds occasionally sing through the silence, and everything is as it should be.

The CDT is obvious as there are no other trails around. No intersections or 4x4 roads. Markers or trail posts are scarce, but not necessarily needed. The dirt trail winds in and out of the shade while the heat of summer releases sweat from my face, pits, and back. The path is dry, and dust explodes from the bottom of my shoes with every step, leaving an imprint of my sole on the earth. I take lunch at a creek and rest in the shade. Dale Creek is the first water source I have come across all day, the last one being way back across the border in Colorado. The creek is flowing strong, and there are a few flat spots that have undoubtedly held tents for a night or tired hikers for a lunch siesta. I am in for a siesta and begin to get comfortable.

The first thing I do is fill my water bottles and place them by my pack. I take off my damp shirt and lie it on a large, flat rock to dry. My hiking shirt holds five different shades of purple, pale from sunlight, and darkened with sweat. I remove my shoes and socks before walking carefully to the creek. I sit on a rock and place my aching feet into the cold, rushing water.

The morning has been hot, and I never pass up an opportunity to soak my feet during a siesta, nor do I miss an opportunity to rinse out my socks, which are so adept at collecting the dust and dirt that is constantly being disturbed by my own feet. I splash water up on my calves. It is easy to tell how high my socks come up. The tan and dust lines combine to color my skin a much darker shade than what hides beneath the protection of a thin wool layer. My socks show signs of deterioration in the toe area. Dirt intrudes the fabric and causes the material to wear out far quicker than it should. Through the deserts of New Mexico, my hiking socks barely lasted two weeks before developing holes. Similar wear is bound to occur as I approach, and walk through, the Basin.

I soak and wring out my socks, one at a time. The first few rinses produce a dark brown liquid resembling pulpy chocolate milk. It runs away with the stream before settling further along. This process takes quite some time and many rinses, but it's well worth it, and my socks feel fuller after a diligent rinsing. I set the wet wool on the rock beside my shirt and leave them to soak up the sun.

This is the best part of my day: right now. The morning is behind me, 11.5 miles walked so far and less remain to hike after this rest. I prefer to do the greater part of my miles in the morning, so I have less to do in the evening. This allows for a longer lunch break and slower walking as I gradually tire with the day. I rest my back against a rock and drink more water.

Pulling out my food bag, I dump what remains on the ground in front of me. I will be in town tomorrow morning, so my food supply is running appropriately low. Eyeballing my options, I return the Ziploc bag of couscous to the food bag and consider the scraps that will be my lunch for now and snacks for the rest of the day. There will be nothing left for breakfast tomorrow.

I normally never run out of food completely, partially due to my habit of buying too much in town and wanting to have plenty to eat, just in case. In case it takes me hours to catch a hitch. In case I come upon another hiker who is much hungrier than I. In case I get really hungry myself. In case I want to spend another night in the woods. Usually, I get to town with a half serving of couscous left, a couple packets of sauce, a handful of nuts, and a mashed granola bar or two. Peanut butter is usually the first thing to go, then the sweets, then the wraps, then the Pop-Tarts.

I lay down in a small patch of grass as the afternoon breeze picks up. I close my eyes and listen to the water move along in the stream as it crashes into itself and tumbles over obstacles. There is a large rock wall protruding beside the creek, creating a tunnel for the wind to blow through. The gust shouts through the corridor, adding to the deafening hum of Nature. It drowns out the silence that has been constant all morning long.

The current background noise allows me to clear my mind more easily than the cadence of my footsteps. I drift into nothingness, arriving at a headspace where time ceases to exist and a place void of thought where I lose all feeling of my body. I am merely a mind, existing in this shell I call my own: a perception, an exoskeleton, a meaningless form.

Maybe I doze off. Maybe I don't. A while later, I take a few deeper breaths, wiggle my fingers, and open my eyes. The shade still covers me, but it has moved to cover my socks and shirt. I rise slowly and move my clothing further into the sunshine. They leave moisture-laden shadows on the vacated rock. I flip my socks and shirt before returning to my backpack. I grab another bar from my food bag and take one small bite at a time, chewing carefully while savoring each flavor that presents itself in the smorgasbord of ingredients. I drink water between each bite, wanting to hydrate as much as possible before starting again.

Trees are becoming more scarce as I walk further away from colorful Colorado. I embrace the change of scenery and am glad to be in a new environment. The deserts of New Mexico left a good, if not sandy, taste in my mouth. I relish the opportunity to see it in another form as the Basin begins to manifest itself here in southern Wyoming. My afternoon siesta lasts a couple of hours, by which time my socks are mostly dry and my legs well-rested. Another eight miles to go for the day. This will take me three hours at the least, four at the most.

The trail immediately climbs up out of the valley holding the creek and forces me promptly into the open. There is not much of a trail, so I move north, making the way as efficiently as I see fit. On top of the short but steep climb, the scenery changes once again. The trees are even more elusive, and the ground glitters in the sunlight. Littered amongst the ground, bits of quartz shine pink and orange. Small flakes blend with the dirt to create sparkling sand that shimmers under the sun. Large chunks peek out from the earth, leaving greater portions iceberged below the surface.

Wooden trail posts disappear and are replaced by sporadic cairns standing up to five feet tall. Composed of rocks and quartz gathered from the ground, they stand spectacularly while blending in well with their surroundings. I grab a bit of quartz from the dirt floor and place it on a cairn for good luck.

The rock statues seem larger than they need to be, yet this is the only landmark around to indicate trail existence and direction. The trail is well beaten in closer to the cairns and indefinite between. I carry on and keep moving, not always able to spot the next cairn. It is not a direct path, nor an efficient one, but I prefer cairn spotting and using my hiker instincts rather than pulling out my phone every time I feel misdirected. The trail is never too far away, and any missteps are quickly corrected. Quartz lights up the ground in front of me. I travel swiftly over the gleaming landscape while the sun grows fonder of the western horizon.

On the back end of summer, the days have been getting shorter. For the most part, this goes unnoticed, but it is a thought that remains in the back of my head. Two states down, roughly halfway, and always getting closer to Canada. The first eighty days of this trip have been more than I could have hoped for. Trees come and go, streams trickle out of sight, but cairns and markers continue to provide direction as I walk north. Every day is the best day ever, and I am grateful to be spending my days walking through Nature.

XII.

SUN, SAGE, AND STORM

July 25 - Day 87

The Great Divide Basin is the exact opposite of the Continental Divide. While the Continental Divide marks a line along the landscape where falling water is divided, eventually ending up in either ocean, the Basin is a black hole where water goes to stay. Any rain that does fall within this perimeter remains to be soaked up by sage, cacti, cows, horses, and hikers, never making it to either ocean. On a map, the line identifying the Great Divide Basin creates a circle in southern Wyoming. The Continental Divide Trail follows the northern part of the circle out of Rawlins for about 120 miles before meeting up with the other side of the circle in South Pass City.

Not all past thru-hikers of the Continental Divide Trail speak very highly of the Basin. After hiking through the hot, dry desert of New Mexico, we were spoiled by the mountains, scenery, population, and frequency of water sources in Colorado. Wyoming's Great Divide Basin promises a return to the days of carrying four liters of water at a time while sweating profusely through the most exposed stretches of desert, unable to stop out of a burning desire to reach the next patch of shade or to find comfort in the next water source. With the Wind River Range looming ahead, the Great Divide Basin feels like a final test before reaching these glorious mountains and an appropriate reward for having crossed the expanse of the Basin.

I left Rawlins yesterday after spending two nights at the Econo Lodge. It was a decent-sized town, but the grocery store was right behind the hotel, which was really all I needed. I did have to walk two miles round trip to the post office to send a couple of resupply boxes ahead but figured it a good opportunity to stretch my legs on a rest day. I iced my ankles all day, watched crappy television, ate decent food, and took an Epsom salt bath to prepare my body for demanding days in the Great Divide Basin.

Water is scarce in the Basin. Reliable sources are triple checked before I take note of the mileage between each one. There are fewer people out here to ask for help than there are decent water sources. A miscalculation can be catastrophic.

Thanks to the limited water sources in New Mexico, I have learned how to pace my thirst and know that I require a liter of water every five miles at the minimum.

I can manage on a liter every eight miles if necessary, but prefer not to. I also smoke less, if at all, during these dry stretches to avoid cottonmouth and any lapse in concentration or common sense. Regardless of the distance between sources, I conserve water and make sure to have at least a liter when I reach the next one, in case it turns out to be dry. Upon reaching the next source, I drink at least a liter before carrying on, to rehydrate after conserving my water over the past twenty miles, and to camel up before carrying on through the next waterless stretch.

This morning I walk a couple of miles beside a paved road before turning onto a dirt road that is perfectly straight for as far as the eye can see. I wave goodbye to the few passing cars on the pavement and take to the dirt, ready to disappear into this great black hole between the lines. I zone out too soon, and within the next hour, I am off the red line. I walk in the wrong direction for a quarter-mile before realizing my mistake. This is not my first time missing a turn and surely won't be the last. Backtracking, I take the correct turn at the formerly misinterpreted junction to follow another mindlessly straight road, further removing me from the highway now well behind me.

I have not seen another soul out here, nor do I expect to. Other than for the reasons only a long-distance hiker can justify, who would walk through the Great Basin voluntarily? It is not an environment catered to day hikers, and traversing the whole Basin in an adequate time frame requires one to hike at least twenty miles a day, which is not something many people can do right out the door. Luckily for me, I am on day 87 and over 1,300 miles into my hike. I can walk twenty miles a day without thinking about it, and I plan to hike at least that until I get to South Pass City.

The openness of this land has me constantly in awe with nothing and no one around for acres. The miles are almost completely flat with a few small, gradual hills scattered about the wide-open expanse. Even in the desert, nothing is absolutely flat. After the steep climbs and descents of Colorado, this terrain is a welcome change of pace. It provides much-needed evidence of accomplishment. The continually changing landscape is confirmation of progress being made. As the days blend together after nearly three months of hiking, the scenery offers a medium with which to measure the impact of my footsteps. These legs are carrying me somewhere, albeit slowly.

It does not rain much in the Basin, but the storm clouds are intimidating nonetheless. Panoramic views allow a clear understanding of the surrounding landscape while offering a show of shapeshifting clouds drifting lazily across the sky. It is still early, but the day warms all too quickly. The weather forecast predicts temperatures in the upper eighties, so I keep my umbrella out to supply my figure with a bit of shade. The slight discomfort in my elbow is worth the protection it provides my torso. My legs continue to tan and kick up dirt in the hot desert sun.

As afternoon approaches, darker clouds begin to weave themselves across the sky, intermingling with stripes of white and strips of gray. I stare at the ever-morphing art hanging effortlessly against a blue backdrop. The flat landscape allows me to see for many miles with no obstruction in my eye line other than the distant curve of the earth. With the sun now hiding behind a vast overlay of fluffy cover, I put away my umbrella and open my arms wide to catch the wind blowing over the desert. There is nothing to block the breeze, nothing to deter it. Evidence of the wind's effort shows above as clouds continue to rush across the sky.

As I walk, I contemplate the continuous movement of everything around me. I am walking roughly 2.5 miles an hour, the earth is spinning 1,000 miles an hour, while rotating around the sun at 67,000 miles an hour. Clouds move across the sky at an unknown rate, encouraged by the strong wind. Everything in sight is moving. Nothing is ever still. Unknown forces encourage constant motion. There is nothing to do but watch.

Our lives are constantly moving, and there is no way to stop the process. Time does not pause. All adjustments and any appreciation must be done along the way. Time does not wait. The proof moves around us always. The sun floats from one horizon to the other every day. Stars drift across the sky as the earth continues to turn beneath old light shining from light-years away.

With this movement all around me, I keep walking to avoid any possible afternoon storms. There is definitely thunder off in the distance and while rain is uncommon in the Basin, it is still possible. I see storm clouds moving across the sky and hear the thunder move with it. Estimating my speed against the pace of the clouds and turn of the earth, I determine it safest to keep moving to avoid any chance storm or freak lightning. As the tallest thing around for miles, walking seems like the safest thing to do. Movement provides comfort while sitting on my ass gets me nowhere.

I keep striding ahead, confident the cloud's path will miss me as I attempt to out-walk the storm approaching from the east. I feel safer in movement as I keep looking around me, anxious to be out of the storm's direction. A few sprinkles spit down from the sky until the roll of the earth brings me out from under the clouds. The dirt road carries on straight while the retreating thunder provides background noise in the otherwise silent basin.

The Basin is full of empty space. The path ahead is not difficult to follow, so my eyes wander across the desert, dissecting the landscape. Dry grass and sagebrush fill the areas between the dirt. No more than a couple of feet tall, the sage subtly radiates a pale green. Its soothing smell is inhaled with every breath. I taste it on my tongue as if the air were flavored. Close to my feet, grasshoppers hop away, lizards scuttle to safety, and birds snap off their perch. The ground jumps to life as I walk through and settles back down with my passing.

***

Toward the end of the day, storm clouds return. This time I am caught under its path, and after cresting over the high point of the day at 7,800 feet, I look around fervently for a place to set up my tent. I do not want to camp on top of this slight hill with a storm approaching, so I dip off the right side of the trail and retreat carefully to a dry wash that offers some protection. I find a flat spot as the air temperature drops around me. The sunset is overshadowed by ominous clouds and, soon, darkness falls.

I set up camp quickly, moving with a speed and efficiency gained from nights upon nights of practice. I stake out the rainfly and duck in for refuge just as the downpour begins. Zipping the fly shut, I leave my backpack in the vestibule just outside the tent. I pull the rest of my belongings in with me while listening to the sky erupt. I sit inside, eating my dinner as the storm that has been threateningly hanging around all day finally unleashes upon the desert.

Raindrops spatter the rainfly while little white pebbles of hail bounce off the taut material. I turn off my headlamp to appreciate the light show projecting itself onto the walls of my tent. They flash white for a moment, overwhelming my night vision. Not even a second goes by before thunder splits open the sky, shaking the ground around me. *I am safe inside this tent*, I think to myself. Wind continues to blow and rattle the rainfly, but the stakes hold firmly in the dirt floor.

The storm must be directly above me. I remain inside the confines of my tent, refusing to even stick my head out to witness the lightning flashing outside my personal haven. *I am safe inside this tent*, I repeat, as sounds of the storm grow louder still. I am thankful for my tent, and this wash I came upon. This is a much better place to wait out a storm than at the top of the hill between two sage bushes. I chose my spot, now all I can do is wait.

This is the most exposed trail I have ever had the pleasure to hike. With no man-made structures or taller vegetation around to provide protection, the forces of Nature must be treated with respect. The element of danger adds a little more excitement to the day. I am happy with the decisions I made today in order to remain safe and alive. With so little out here, these choices hold more weight than in places where refuge can be taken under a grove of trees or in the protection of an outhouse. One wrong decision could make for an uncomfortable afternoon or uneasy evening.

I return to my dinner as the storm continues to pass overhead. Bright flashes of lightning and explosions of deafening thunder continue to startle me. After finishing dinner, I lie down on my sleeping pad and close my eyes to the show around me. *I am safe inside this tent.* The orchestra of the Basin continues to play its thumping music as I drift into sleep. A full marathon through the Great Divide Basin has granted me with scenes seen nowhere else. The mystery and menace this place contains are unrivaled by any of the enchanting deserts I walked through in New Mexico. Although this area does not receive much rainfall, these short, heavily tempered storms move in quickly and linger a bit before dissipating.

These are the experiences that make my blood pump and remind me how thrilling life can be. As a human, I crave challenges that will carve me into a better, more capable person, and what challenge can be more rewarding than survival? With exposed trail and limited water sources, the margin for error is small. I carry only a backpack yet still feel capable of handling anything this wild, raw country can throw at me. Eighty-seven days into this adventure, and my confidence grows every day. I am thankful for these simplicities amongst Nature while choosing to forego the luxuries and superfluous clutter of modern life. With less, I find it easier to sing along with Nature's chorus.

In the middle of Wyoming
Neither walking to nor going,
There is no way of knowing
What kind of walk walks through Wyoming.

COLD SOAKING: A DAY IN THE LIFE

Cold soaking is the culinary art of backpacking without a stove (see also, *stoveless*). While most hikers carry a stove to cook meals with (mainly just to boil water), some choose to forgo this luxury. This saves weight and space in your pack while simplifying meals on trail and resupplies in town. My cold-soaking container is an empty 24-ounce peanut butter jar. Accompanied with a long spoon, this universal bowl is all a stoveless hiker needs in his or her kitchen.

Breakfast

The first meal of the day. The most important meal of the day. Two bags of oatmeal and a sprinkle of whatever else may be in my food bag: chocolate chips, dried fruit, granola, a third bag of oats. I prepare this the night before, so in the morning all I have to do is add water and start walking. A hot tea or cuppa joe can be enjoyed at the local coffee shop in town. Walking has never failed in warming or waking me up, and I will continue to rely on this technique to start my days.

Lunch

Lunch is usually a collection of snacks. I have been known to throw a bar in a wrap with miscellaneous toppings. A Pop-Tart is commonly consumed around this time. My favorite flavors in no particular order are cookies & cream, s'mores, strawberry, blueberry, wild berry, and brown sugar. Other common snackings include, but are not limited to, dried fruit, Triscuits, Fritos, potato chips, chocolate chips, granola, granola bars, protein bars. It's all more or less picked at.

Dinner

The main reason most hikers choose to carry a stove is to enjoy a well-earned, hot meal at the end of a long day full of ups and downs. Meh. At the end of the day, I just want food and nourishment, who cares if it's hot. Spoiler: it's not. Sometimes I buy ramen for economy's sake, but for the last month, I have been all aboard the couscous train. Cous-cous!

Just add water to this strange cereal of a pasta and watch as the crunchy little nuggets grow twice their size, softening in the process. Or put the bowl back in your pack and hike on. Dinner will be ready whenever you decide to drop your bag and set up camp for the night. One tortilla is always used to make a burrito, adding sauces and spices accordingly. Like Grandma Gatewood used to say, "Everything tastes better in a wrap!"

Tips

- A deep selection of snacks is important for any thru-hiker. Be prepared.
- ABS. Always Be Saucin'. Check gas stations and grocery store delis for packets.
- Pack out fruits and vegetables to enjoy the first couple days out of town.
- Make sure to eat a real, hot meal in when in town. If not two.

XIII.

DO IT WHILE YOU'RE YOUNG

August 2 - Day 95

Pinedale is a resupply point out of the Wind River Range. Some hikers skip this town, preferring to hitch into Lander prior to entering the Winds. This requires hiking bigger miles or carrying a larger resupply in order to make it through the mountain range in one shot. I didn't go into Lander and do not mind hiking the extra miles to reach Elkhart Park Trailhead, where I can get a ride into Pinedale. Yesterday evening I departed the red line and began hiking the Pole Creek Trail, which will lead me to the trailhead. I stayed the night at the northern end of Chain Lakes, a few miles into the alternate.

This morning is wet thanks to a heavy storm that came through yesterday evening. Although I begin the morning dry, my jacket and leggings are soon soaked from walking through brush that is overhanging on trail. It is 5:30 a.m., and the sun will not rise above the mountain tops for another couple of hours. I curse last night's storm and continue through the thicket with heavy feet.

The sun eventually shows itself between the horizon and a cluster of low hanging clouds. I stop beside the trail, set up my tent, and hang my rainfly in a tree. Allowing the breeze and dull morning sun to dry my things, I enjoy a second breakfast and rest. The sun lingers for a moment above distant mountains before playing a game of hide-and-seek between wisps of clouds. The slight breeze carries through my damp gear as I watch water dry. If there is rain the night prior, appropriate time must be spent drying out all wet gear as soon as possible. Wet clothes and gear can develop mold if bundled up and ignored. Also, a rainfly soaked with water is heavier than a dry one.

I pack up my gear once it's dry and stuff it all into my backpack. This alternate adds a few more miles since I am going off the red line to reach Elkhart Park Trailhead, but I don't mind at all. The miles are easy, and I get to see more of the Wind River Range. North of here, the CDT will continue to wind up and over incredible passes and past breathtaking alpine lakes, so I am pleased to be seeing another side of this beautiful wilderness.

On a thru-hike, you do just that. The trail is not always direct and is sometimes the most challenging route from A to B. The Continental Divide is not a straight line from border to border, so you can't expect the trail to be. The objective is to go through and be a walking witness to what is along and within sight of the trail. This alternate provides another perspective to a wilderness that is so much more than its high peaks and demanding passes.

A few miles out from the trailhead, I catch up to a fellow hiker. I evaluate him from behind and guess he is headed back to his car at the trailhead. I am going to need a ride into town once I get to the trailhead, so when he steps aside to let me pass, I slow my stride, allowing him to keep up with me. Conversation is easy to start on trail and easier to keep on. If a person is going backpacking, they are alright in my book. At least they're trying. This guy's name is Drew, and he looks to be around my age. He and a couple of other supervisors took a group of boy scouts out to camp for a week, but he left a day early to make it to his sister's wedding this weekend.

"How long have you been out for?" he asks.

"Ninety-five days."

"Oh shit, you're doing the CDT?! What are you doing down here? I thought the trail was further east."

"It is. I'm just going into Pinedale for a resupply."

"How are you getting there? Do you need a ride?"

"Yes, please! Are you headed that way?"

"Yeah! I'm goin' back home, so I'll be driving through there anyways."

Conversation carries us down the trail. I do my best to slow down even more because I sense he does not want to leave the woods just yet. Whenever I tell someone I am hiking the CDT, they usually react in one of three ways. Most think I'm crazy: "I would never be able to do that," they reply out of disbelief. The other third acknowledges the feat: "Good for you," is usually the response of those who hold thru-hikers under a brighter light, willing to measure us with a different ruler. The final third is the saddest and mopiest of the lot: "I wish I could do that," is the response spoken from this envious, Eeyoric bunch. This response frustrates me because their wishes will burn them up from the inside. I despise myself for bringing out this emotion in others.

At least the first group is honest, and the second is respectful. The latter is just frustrated, stuck spinning tires in the mud of their own resentment. Drew is in this group. I want to bring out the best in people, yet these brief moments of self-inflection from participants stuck in the cog can be discouraging. They quietly question past life choices that have led them to this point and wonder how they evolved to be the well-established societal contributor offering a ride to a thru-hiker when they would much rather be the one asking for it.

The more we talk, the more we learn about each other. We are close in age and get along easily. I am just as interested in his life as he is in mine. I don't like to always be the one answering questions, and part of this experience is about meeting as many people from various walks of life as I can. This country is full of interesting people, and you can learn something from everyone. The world and its inhabitants are merely a walking library, waiting to be flipped through and analyzed.

Drew, just like I, went to college for four years and got a degree. He went on to begin a career in engineering, got married, and started his family, which is a very admirable path to choose and not one to be ashamed of. I took the route he did not and the route many in his situation wish to be more familiar with. Out of college, I hiked the Appalachian Trail and then began seasonal work as a ski instructor. Four years later and I am still living seasonally, continually figuring out the next step while setting my foot down on what I hope to be sturdy ground. We walk different paths in life, but one is never to be better or more justified than the other.

His envy and longing are only to that which is foreign to him. It is not so much my lifestyle he envies as much as my freedom. With a wife, child, car, and mortgage, his freedom is limited due to these commitments that require him to spend a majority of his hours working and taking care of his family.

Things may open up for him twenty years down the road but, this wait will torment his mental state as he sits behind his desk pushing buttons and measuring lines on AutoCAD.

All too soon, we make it to the trailhead.

"Do you mind if we take a picture together?" Drew asks unabashedly.

"Not at all."

No one is around to take our picture. He holds the phone in front of our faces as we hang our arms off each other's shoulders. We are all brothers of the woods and being able to share even a few miles with someone is enough to get to know them well enough to exchange numbers or capture the moment. This picture will be his proof of having met and walked with a thru-hiker of the CDT. Proof to his family and friends that it is possible to live a life in the woods, jobless, for six months. Proof that you do not need a high-paying job or white picket fence to be happy. Joy is found in the simplest places, joy is felt in the woods, joy is brotherhood, and joy is effortless with a pack on your back.

In the car, he reflects on his hike before asking more about my journey. It is easy to recognize he would much rather resupply with me and return to trail than go to his sister's wedding. There are no worries in the woods, no plans, and no stress. His five-hour drive back home is sure to be infiltrated by playful thoughts of what it would be like to return to the wilderness to walk for another few months. We descend from the Wind River Range and roll down the road to lower elevation. The ride to town is full of conversation still, and I feel his sadness for having to leave Nature behind for now. Compared to my ninety-five days on trail, his five nights spent with a group of boy scouts sounds feeble. He longs to be back on the trail, and I can feel it. Jealousy radiates from his words as I try to downplay how far I have come.

Other's envy of this life provides another dimension to my justification of spending half a year walking border to border. If they envy my actions, then it must be better than the alternative, right? The views I am afforded are not going to be attained by sitting behind a desk, and they certainly cannot be bought with any amount of money. Sacrifices I make in the present day may come back to haunt me later in life, but the lure of the outdoors is far greater than any office, cubicle, paycheck, or promise of promotion.

I imagine folks sitting behind their computer at work or else scrolling through their phones while on the couch, reading my blog posts and seeing my pictures. At this point, I crave their envy because I know it may inspire them to do something that makes them happy in an unusual or uncommon sort of way. My removal from the typical work year can prove to others that it is possible to only work six months out of the year in order to enjoy life a little more.

While on a thru-hike, I am reminded the path is laid out in front of me, and all I have to do is walk it. Connections and lessons learned from the past have led me to where I am today, and as long as I let the world turn as it does, things will continue to happen, most of which are bound to be in my favor. My faith remains in the universe, grounded in Nature and encouraged by chance. The trail provides, and the same can be said about life. As long as you get up every day and move a little further along than you were before, you will continue improving and be allowed to see the magic behind the magnificent insignificance of everyday life.

Drew drops me off at the grocery store in Pinedale. I say goodbye, take my backpack from the trunk of his Prius and transfer it to a shopping cart. Inside, my stench carries two aisles over as I stroll through the store, enjoying the air conditioning while dumping bars, pastries, and snacks into the rolling food carrier. Resupplies have become automatic. I know what I like, and I know what I don't. I know how much food I need to eat each day, and I know how many days of food I need.

Eighty-nine trail miles to the next road crossing where I will be hitching into Dubois. Eighty-nine divided by twenty is roughly four and a half days. My mileage increased slightly across the flatland of the Great Basin, but back in the mountains, I plan to average around twenty miles a day. There will be many sights to see and lakes to soak in. The passes are going to be steep, and I want to take my time through these beautiful mountains. This past stretch from South Pass City to Elkhart Park Trailhead was just a tease of what the Wind River Range has to offer, and the best is always ahead.

The night prior to arriving in town, I make a to-do list and know what I have to do before getting back on trail. Top priority is to resupply on food for the next stretch. This has been roughly estimated the previous evening, so I know how many days' worth of food I need. Food can be calculated by meal type (breakfast, lunch, dinner, and snacks) or by calories (3,000-4,000 per day). I have done both but prefer to calculate by meals because counting calories does not always equate to sufficient nourishment.

Some of the upcoming towns in Idaho and Montana are pretty remote, so I also need to pick up an additional resupply which will be sent forward to Leadore, Idaho. On trail, I never like to think too far ahead, but occasionally, a resupply box has to be sent out a couple of weeks before my arrival in these towns that do not have a proper grocery store or fairly priced general store.

All throughout Ridley's Market, I receive weird glances. People turn around or adjust their paths when they see me approaching. This treatment is normal, but it reminds me to take a shower while in town. Just because I am oblivious to my smell does not mean everyone else is. I stare at the jumble of boxes in my cart, counting out calories and meals. It looks like enough food, and I can always come back in to buy more if needed. Before heading to the checkout line, I grab a bag of bagels, a container of hummus, a cucumber, and a pint of non-dairy ice cream for lunch.

The lady behind the register starts scanning my stuff as I set it on the conveyor. She notices my backpack in the cart.

"Where ya hikin'?"

"I'm on the CDT."

"What's that?"

"Continental Divide Trail. Goes from Mexico to Canada."

"You doin' the whole thing?"

"That's the plan."

"Is it hot out there?"

"Nah, it's not too bad. It's hotter here in town."

It's true. Buildings, concrete, and asphalt generally make the air at least ten degrees warmer. There are no trees, dirt, or grass around to properly absorb the sun's heat or provide natural shade and protection. The hard sidewalk is unforgiving, and stale heat collects on artificial surfaces that continue to release warmth from underneath.

I insert my card into the card reader as a hundred dollars is removed from my funds. Two resupplies and lunch for a hundred dollars, not bad. Most of the time, my eating cost on trail comes out to ten dollars a day. I can eat for a full day on ten dollars and hike twenty miles in the woods, or I can go to the bar, have two beers, and still have ten hours left in the day to fill with more nonsense or spending of money. Food is cheap, society is expensive.

Outside the grocery store, I sit on a couple of empty pallets tucked up against a soda vending machine. There is an outlet nearby which I plug my phone into. I withdraw a dollar from my wallet and insert it into the vending machine for a pineapple pop to enjoy with my ice cream.

I dump my bags of food on the ground and begin to remove the individually plastic wrapped items from their cardboard shells. Cardboard goes into one bag while trash goes into another. Four and a half days' worth of plastic-wrapped goods and topped off Ziplocs go into my food bag. Another six days' worth of nourishment is placed into a couple of plastic bags so I can carry them to the post office, which is half a mile from where I am now.

I make it to the post office as a collection of clouds moves in for an afternoon storm. I stuff six days' worth of food into a priority express box and pay the flat shipping rate for it to be sent to Leadore. Six days is about the maximum amount of food I can fit in the large box that costs $13.95. If it fits, it ships! Snacks and bags are strategically placed like Tetris blocks to maximize every square-inch of space inside the box.

The package weighs fourteen pounds, and I dread the day I have to pick it up farther north to carry this weight on my back. I check another box on my town to-do list and look up where the aquatic center is so I can spoil myself with a shower. Another half-mile. Town miles are tedious but necessary. This is not even that large of a town, but to a hiker, if it's not across the street or next door, it's too far.

I pay to take a shower at the aquatics center before returning to the lobby to eat the second lunch of bagels, hummus, and cucumber I bought at the grocery store. I plug my external battery into one of the many wall outlets and return to the table to eat and look for new hiking shoes online.

A couple of ladies sit at a nearby table and eye me suspiciously. They steal glances at my backpack, taking note of my tan skin and thin legs while their children run aimlessly around the lobby.

I meet eyes with one of them and smile, "Hello! How are you this afternoon?"

"Good!" she replies. "Are you hiking somewhere?"

"Yeah, I'm on the CDT. Going through the Wind River Range right now."

"What's the CDT?"

"Continental Divide Trail. Goes from Mexico to Canada."

"How much of it are you doing?"

"The whole thing."

"The whole thing? How long is that gonna take you?"

"Five or six months. Today is day ninety-five."

"Oh my goodness, you've been hiking for ninety-five days? What do you do for food? Which way are you going?"

"I stop in town every four to six days. I'm going north. Started at the Mexico border."

"Wow, that's incredible. How are you able to do this? Do you work?"

"I worked two jobs throughout the winter and saved up so I could take six months off to do this."

"Good for you. You don't have any kids?"

"Not that I know of."

"Well, that is all very impressive. Isn't it dangerous? Have you seen any bears?"

"There's no place I feel safer than in the woods. No bears yet, but I am gonna pick up some bear spray before getting back on trail."

"That's a good idea. What all do you carry? Your backpack is tiny!"

"Tent, sleeping pad, quilt, food, jacket, and a change of clothes."

"That's it?"

"Pretty much, I've got a small bag of toiletries, an external battery, headlamp, journal. I only take what I need."

"I couldn't do that."

"It's just walking."

"Yeah, but I have these kids to take care of," she says, looking around to locate her offspring running amongst the throng.

"I couldn't do that," I admit.

"Well, do it while you're young. Once you have kids, things get a little complicated."

"So I've heard."

"Sorry, I have one more question."

"Shoot."

"Do you call your mom?"

I laugh before responding, "Yes, of course. Whenever I'm in town I call or text her so she knows I'm alive and well. Thank you for reminding me though, I'll do that once I leave here."

"Well, good luck. Stay safe out there."

"Thanks, I will. It was nice talkin' to you. Take care."

Whenever people tell me, "Do it while you're young," I laugh inside. What does it look like I'm doing? And I can only do this while I'm young? Then what? Then it's all downhill once you are no longer young? If I can only do this while I am young, then what is the point of getting older? The people who tell me this speak with bitterness in their voice and envy in their eyes, neither of which they are capable of hiding.

Do it while I'm young? I'm doing this for as long as I can.

XIV.

SHINY NEW TOY

August 10 - Day 103

A truck starts and pulls out of the gravel parking lot. I stir in bed as a low sun sneaks its way through a gap in the stale curtains, projecting a strip of light over the sheets. Covers lay twisted between my legs as I groan to the ceiling and roll over to check the time on my phone. The 7:15 alarm is still twenty minutes from singing its startling ring of repetition.

I run tanned hands over my face and attempt to recall the dream that caused me such a restless sleep. It was another subconscious mind ramble involving my old campervan. This time I was replacing a part in the engine while parked in the driveway of my childhood home. You figure that one out. I pull my legs to the side of the bed and hope for strength. I walk a couple of steps before a pain shoots through my left ankle.

"Augh!"

I drop back onto the firm mattress. The past two mornings were spent doing the same thing. I take a breath, regain my feet, and walk gently to the toilet to relieve myself.

Not bothering to flush, I turn out of the bathroom to look upon my belongings scattered about the room. Camping gear and clothes cover much of the dark orange carpet, hiding stains from the nineties. A sun-faded backpack is doubled over in the corner, without a frame, it folds helplessly in on itself. This motel room has been home for the past two nights, the past two days of which were spent icing my sore ankle while sitting outside in a plastic Adirondack chair. During the hotter hours of the afternoon, I retreated into the room to watch reruns of *Friends* on the box television.

I step softly to the microwave where my wallet is. Unzipping the small pouch, I withdraw a business card that was given to me two days prior by a welcoming local. A small picture on the otherwise white card shows a man smiling behind sunglasses, holding a healthy-sized cutthroat trout in his hands. Gary gave me a ride into town and told me to call him if I needed anything.

This is my second time calling within the past two days, our prior conversation a precursor to this one. While typing his cell number in, I walk outside to get better service. The phone rings twice before it is answered on the other end.

"Mornin', Brian."

"Mornin', Gary. How are you today?"

"Good, good. How are you?"

Awful. Just awful.

"I've been better. Could you give me a ride to the clinic this morning? My ankle isn't feeling any better, and it's about time I get it checked out."

"Sure, I've gotta drive down to Lander later today, but my morning's free. When do you want to go?"

"Is 8:00 too early? I wanna be there when they open since I don't have an appointment."

"Yeah, that's fine. Are you still at the church?"

Churches often allow hikers to sleep in their basement or conference room for a small fee/donation. I spent my first night in town at the Episcopal but checked in to the motel so I could sleep on a real bed and be closer to the grocery store.

"No, I slid over to the Wind River Motel."

"Sounds good, I'll come get you around 7:50."

"Thanks, Gary, I really appreciate it." I try to keep my voice steady while holding back tears of gratitude and distress.

"No problem. See you soon."

"Will do. Bye."

When in town, I usually walk wherever I need to go. Unfortunately, this injury has made a three-mile round trip seem inconceivable, a thought that brought tears to my eyes the day before as I considered the thousand miles of trail that remain between this town and the Canadian border. Today, all that is being pushed to the back of my mind. One day at a time.

Retreating to room number twelve, I pour Reese's Puffs into my cold-soak container and splash the crunchy spheres with almond milk. The 7:15 alarm goes off unnecessarily as I sit somewhat awake, looking out the open door.

When Gary pulls in, I am already sitting outside, pondering the day's possibilities. Limping over, I hop into the passenger seat and return his greeting. We make our way across town to the health clinic as he paints his town with words. He points out certain buildings: the best breakfast spot downtown, the church he goes to every Sunday, and the high school where he used to be a principal and an art teacher (not at the same time). Today he is a trail angel.

"Let me know when you're done, and I'll come get ya," he says, pulling his truck into a parking space in front of the clinic.

"You got it. Thanks, Gary," I reply, stepping out carefully.

It is still before 8:00, so I sit on the bench outside, waiting for the door to be unlocked. I sit with my thoughts, reinforcing positivity, and hoping for the best diagnosis possible. Soon enough, a lady turns the deadbolt, opens the door, and welcomes me inside. At the front desk, I'm handed a clipboard with a few forms to fill out.

"I don't have any health insurance. Do you offer any financial assistance?" I ask the secretary. Having turned twenty-six only two months prior, the coverage I was being provided under my parents' policy booted me off the insurance plan.

"Yes, we do," she replies. "We have a sliding scale based on your income. Just write down how much money you've made this year on this piece of paper, and we'll use that to adjust the cost."

I take the clipboard and sit down to fill out the papers. Home address? Don't really have one. Homeless Status? Still trying to figure this one out. Occupation? Nope. Yearly income? I've only worked three and a half months so far this year. When I return the clipboard to the secretary, a nurse takes me back to get my vitals checked out. After hiking over 1,600 miles on the Continental Divide Trail, my resting heart rate is impressively low, clocking in at just 43 beats a minute.

I answer a few more questions and the nurse leads me to the x-ray room where she tells me to lay down on the exam table while she prepares the machine. She takes the lead apron off the wall and places it over my body. Three x-rays of my left ankle are taken and developed before the doctor comes in to see me. Shoulder length gray hair trickles out of a hairline that doesn't begin until the crest of his head. He looks like the old scientist from Jurassic Park and introduces himself as Dr. Johnson.

"What'd did you do to your ankle?"

Where to start?

"Well," I begin, "I walked here from Mexico on the Continental Divide Trail, and earlier this week, the bottom of my left foot began hurting a lot. I woke the next morning with my ankle very stiff, making it extremely difficult to walk. It took five miles of limping along in serious pain before I could even move my ankle at all and another five to have somewhat normal movement. I've taken two days off here in town, icing on and off all day long. It hasn't gotten much better, so I figured it's time to get it checked out."

My personal diagnosis is that it's an overuse injury: a result of walking an average of sixteen miles a day for three months.

"Sounds like an overuse injury to me," he says, pulling up my x-rays on the computer and looking over them one at a time. "I don't see any fractures in your foot or any bone fragments floating around, so there's nothing broken."

A huge wave of relief sweeps over me. For how much my ankle was hurting the past few days, I would not have been surprised if something was broken. I still managed twenty miles a day out of necessity but struggled to hike them in a timely, efficient, or jolly manner.

"The spacing between your bones looks good," he continues, "but since this is only an x-ray, I can't tell you anything definite about your ligaments. However, it is likely they've been worn down and are just weak. Rolling your ankles repeatedly can cause minor sprains, which deteriorate these ligaments over time. Without adequate recovery time between sprains, your ligaments will only stretch and tear more easily."

"So then the pain in my foot is just a consequence of the overstretched ligaments in my ankle?"

"Correct. The lower ligaments in your foot are working harder to make up for the strength lacking in your ankle, which is causing the pain to drop lower in your body."

The knee ligament is connected to the ankle ligament. The ankle ligament is connected to the foot ligament. Gravity is an incredible thing. But what does this mean for my hike?

"Can I keep hiking?" I ask with a slight tone of desperation.

"We call this a chronic ankle sprain. Since there's nothing broken, you can keep hiking on it, as long as you can tolerate the pain. That being said, I recommend taking a week off and cutting your mileage in half."

Half??

"I will also give you an ankle brace to wear at all times," the doctor continues. "Make sure you are taking time to stretch every day and after a couple of weeks you can do some exercises to begin rehabbing the ligaments while you're hiking."

How much is that ankle brace gonna cost, Doc? I think to myself, but instead speak aloud, "So it's not going to get any worse if I keep hiking on it?"

"That's right. Take ibuprofen for the pain if you want, it'll also help with the inflammation. I'll go get that brace."

After Dr. Johnson leaves the room, I hop off the examination table and step toward the computer to examine the pictures of my insides. I consider his diagnosis and resulting recommendation. Nothing broken and not getting any worse. That's the best news I have heard in a while.

The doctor returns with the ankle apparatus and shows me how to use the velcro straps to adjust the tightness. Thanks, Doc, I can take it from here. After shoving a slightly wider ankle into my left shoe, I shake his hand and thank him for his time. He wishes me luck and tells me to be careful.

Out in the waiting room, the secretary informs me that I qualify to be in the lowest bracket of the pay scale, and therefore only responsible for twenty percent of the total cost. She swipes my card for a twenty-dollar copay and tells me I'll be receiving a bill in the mail. Maybe not having health insurance isn't so bad after all. I leave the clinic with a slight smile on my face and a shiny new brace wrapped around my ankle. It feels bulky, yet supported.

Back at the motel, I go into the small lobby and pay for another night. I will take a third zero in Dubois and spend the rest of the day contemplating my options. Chronic ankle sprain is just a fancy word for weak ankles. I am reminded of my college soccer days and of going into the training room every day after practice to rehab my ankles. You would think walking 1,600 miles is a great way to rehab anything, but I guess I roll my ankles a bit too often for the relentless rehabbing to remain effective.

The trail has been demanding up until this point, and not every mile has been on smooth, even tread. Especially on top of ridgelines, it is common to be billy-goating on steep hillsides or scrambling across inconsistent rock fields. There is also plenty of cross country travel where no trail exists, forcing my feet to constantly adjust to uneven ground, always risking a roll of either ankle.

What Dr. Johnson failed to diagnose, however, is my case of border fever. Border fever is caused by a hiker's desire to get out of the state he or she is currently in and closer to the international border looming in the distance. A hiker with border fever will end up increasing mileage to get to the next state border as soon as possible. The resulting border crossing will only temporarily alleviate the symptoms, which are bound to recur when approaching the next state.

Side effects of border fever may include tired or achy legs, substantial increase in appetite, tunnel vision, hanger, bonking, slimmer legs, depression, sudden increase in morale, changes in mood, and talking to yourself. Take ibuprofen, meditate, walk, or smoke marijuana to calm these effects. If symptoms last longer than two weeks, keep hiking. You must be close.

Many cases of border fever are derivative of the even more severe maple fever. At this point, I am on the bubble of whether or not I will make it to Canada before the snow falls. I was upping my mileage through Wyoming because I felt strong and kept thinking about that northern terminus. I still contemplate a flip-flop hike (going up to the Canadian border and hiking south from there), but I am especially keen on finishing this journey in Glacier National Park: the Crown of the Continent. This is not a new sickness, but in fact, one I have been suffering from since six months before starting this hike. It comes in waves.

Things can be done to combat the sickness, but often, the sickness feels so good that you don't want to. You refuse to fight it, instead allowing it to take over your motivation and fuel your endeavors. You forget what made you so sick in the first place but know that getting to Canada will temporarily cure the disease until the terminal bug infects you again. You turn the sickness into your medicine, and it, in turn, becomes your strength. It gives you a reason to get up in the morning and hike. You learn to live with it and strive because of it. Half of me desires to rectify the disease while my other half wishes to indulge the disorder.

Half. Cut my regular mileage in half. Instead of walking twentyish miles a day, the doctor recommends I only do tenish a day. Halving my mileage also means doubling how long it takes for me to get anywhere. An eighty-mile resupply would take seven or eight days to complete as opposed to three or four.

Further north in Idaho and Montana, there are 120-mile stretches between resupplies. My ankle will have to be feeling a lot better by then, or else I will be in serious trouble. Fortunately, the resupplies in the nearer future are all a lot shorter. I can get away with only doing ten to fifteen miles a day and carrying five or six days of food. For now.

The only thing that matters at this moment is getting back on the CDT and making it to Yellowstone. There is no need to think about the longer resupplies in Montana because I am in Wyoming. Still in Wyoming. A thru-hike is just a bunch of consecutive section hikes. Focus on one section at a time, and do not worry about the future. What I can control is here and now.

XV.

NATIONAL PARK ZOO

August 16 - Day 109

Upon entering the park from the southern boundary, I am greeted with wide sweeping valleys. Trees litter the hillside as a stream winds through the low point of the valley, casting a green gradient across the landscape. Fish swim through the water and flop beyond the mirror-like surface as sunlight reflects off their tiny scales. Patches of burn areas leave temporary scars on the yellow-green hills. Trees left by fires of years past lay helplessly on the ground or else stand naked in the high summer sun. I look around for signs of wildlife but am more attracted to the butterflies flitting from flower to flower in search of the yellow powder that keeps them going. Cast your gaze too far ahead, and you are bound to miss what is right in front of you.

Twelve miles hiked today. This will be my second night spent in Yellowstone National Park. The miles have been extremely easy and manageable over the past few days. Walking half my usual mileage, I am much less tired at the end of the day, and I also have a lot more time on my hands to do things other than walking and eating.

Before this ankle injury, I was averaging twenty to twenty-six miles a day, walking anywhere between two to three miles per hour. I would walk ten to twelve hours a day, which took up most of the day's hours. Halving my mileage halves the amount of time I spend walking, which has opened my day up quite a bit. I am forced to sit a lot more and spend my time in other ways.

Fortunately, I found *The Autobiography of Benjamin Franklin* at the thrift store in Dubois for fifty cents, which is possibly the best investment I have made all year. Much of my time resting alongside the trail is precursed by me locating a good bit of shade with a decent seat so I can read while protected from the sun. I write in my journal more and sometimes just meditate while watching tree branches sway to the breeze.

What I recognize most during this stint of walking half my normal pace, is how much time is in the day. If you do something for twelve hours a day, you are bound to get pretty good at whatever it is you're doing. I realize how much time you put into something determines how good you are going to be at that thing. If you go to the park to shoot hoops for twelve hours a day, you will get pretty good at shooting hoops. If you spend twelve hours a day sewing, you will get pretty good at sewing, becoming a more efficient and adept sewer over time. Repetition and practice do not ensure perfection, but it certainly is crucial to develop a skill.

However, it is difficult to do something for twelve hours a day, and it can be overwhelming to have this much free time to yourself. It takes discipline to focus your attention on one thing for such a long period. Even walking can get repetitive and become exhausting, the same motions being made with every other step. This kind of work is physical, but the discipline to do one thing for twelve straight hours is a constant mental battle.

In the other world, we do not usually have twelve hours in a day to do whatever we want to. Weekends and days off are usually filled by a variety of activities, not just the repetition of a single one. Most people only work eight hours a day because, after that, the mind starts to lose focus, motivation, and attentiveness. During my last year in Mammoth Lakes, I worked production at a brewery where ten-hour shifts were standard. After a couple of months, I adjusted to the longer workdays, but if the day went any longer, my mind and body would begin to fade. Although we work this many hours a day, we are only encouraged to do so because of monetary compensation and because someone is telling us what to do with this time.

Three and a half months ago, I was standing at the Mexico border. Walking and fueling my body is how I have been spending my time since then. This has gotten me in really good shape to walk and has carried me through New Mexico, Colorado, and most of Wyoming. The more you do something, the better you get at it. The more time you spend doing it, the more efficient you become. In this way, time allows us the ability to learn and develop ourselves in whatever way we choose. What do you want to spend your time doing, and is it fulfilling for you? Time spent working toward something is time well spent. Looking back at the end of the day and measuring what you have done with the hours you were awake provides purpose.

Every day I move more toward Canada and closer to my goal. Some days I ponder the distance remaining and calculate an average pace in order to finish by October 1st, which is usually around when the first snow of the season falls in northern Montana. At this point in my hike, I do not know if, or when, I will make it to Canada but keep telling myself not to worry about it. Every day spent hiking, soaking up the day's rays, and every night spent sleeping under the stars is the best way to spend time. Part of my brain and ego desires and expects to make it to Canada because otherwise the point of hiking from Mexico to Canada will be lost. Another part of me is perfectly happy with carrying on at whatever pace seems right, taking this extensive trail one resupply at a time.

It can be difficult to wrap your head around hiking for 6 months and traveling over 2,500 miles on foot. Reading a 2,500-page book can be a daunting task, but if you read a little bit of it every day, the story will be over eventually. A 2,500-mile walk is the same way. Move every day, and progress will be made. If you do it for twelve hours a day, day after day, you will soon enough find yourself at the Canadian border thinking about how you walked all that way from Mexico when all you really did was spend your days walking. Apply this amount of time, effort, and planning to anything, and the possibilities are endless.

I am well into the latter half of my hike, and though I don't want to think about it too much, I must spend appropriate time pondering the future. When I left Mammoth Lakes in the spring, I held the intent of moving somewhere else after the hike, to experience new challenges and encourage growth in different areas to become a more well-rounded human. A thru-hike is not going to solve all of your problems, if any, but what it does provide is a space void of distractions. This allows me adequate time and mental clarity to properly analyze my current life situation.

After enough contemplation, I can make the best possible decision for me. Life problems always exist, and it is a constant game of solving a bit of the equation at a time. No one has all of the answers, but it is good to develop yourself as a person in order to at least ask the right questions before considering a temporary solution.

One of the benefits of thru-hiking is that you are able to spend so much time in your head. You have freedom to reflect on the past, cherish the present, and consider the future. While it is never good to dwell on the past, my mind notices patterns that have encouraged past life choices and particular events that have shaped me to become who I am today. The past is just that, but if you make mistakes without learning from them, then it is easier for you to repeat those mistakes. All too soon, they become habits ingrained in your living code.

I spend a good amount of time on trail pondering my future: what I want out of life, what I want to accomplish, what challenges I want to face, what I want to look back on in twenty years and think, *I am sure as hell glad I did that.* This trail has already taught me that anything can be done, and all that is necessary to accomplish most things is time.

Time is all we have. We have no way of knowing our limit on life yet must spend it wisely, even though such an ability is only attainable through experience. It is what we do with our time that allows us to feel worth or value in life. Time is a gift to all that is always being presented. A gift that must be spent right away, wisely, and willingly.

***

August 20 - Day 113

National Parks protect a chunk of the wilderness, allowing visitors to see a more delicate and raw side of Nature, but a certain portion of wildness and mystery is sacrificed in order to educate and demonstrate. The whole park may have a complete trail system to be traveled by foot, but most visitors stick to the asphalt path that winds through the park, offering just a glimpse of what is out there. Roadside stops may allow for a great picture, but the true riddle of Nature is too elusive to be read this way.

Going through a National Park while hiking such a desolate and unknown trail is the closest thing to culture shock a thru-hiker can experience on the CDT. A couple of days after leaving Grant Village, I walk into Old Faithful Village, excited to see the show and be surrounded by people.

Before entering the Village, a part of me was looking forward to the popularity of the area. After a couple of hours in Old Faithful Village, a larger part of me desires to get through the mass congregation of people as quickly as possible so I can get back to having the whole trail to myself.

Old Faithful Geyser goes off multiple times a day, roughly every ninety minutes. The estimated time of each eruption is posted all around the Visitor's Center. They can predict the next one depending on the time and duration of the one that precedes it. I get to the viewing area early to get a good seat for Nature's display. The benches gradually fill up as the predicted time approaches. Surrounding onlookers speak of the roads that brought them here, how long this has been on their bucket list for, and how many days they spent driving to get here. I sit and write in my journal, not wanting to brag about my walk and not wanting to answer the same questions or to be looked at in disbelief. I already feel alone enough without being gawked at like an exhibit.

Anticipation grows as expectations continue to build. Everyone is waiting with their camera at the ready, anxious to capture the moment for themselves and to share with the internet later, an image that has been widely seen by billions already. The estimated time of eruption comes and goes.

"It's late."

"They're wrong."

They are trying to predict when water shoots out of the earth, give them a break. I wait patiently as those around me fidget in their seats and check the time. Only a few minutes after the predicted time of eruption, water shoots a hundred feet into the sky, followed by a thousand resounding camera clicks. The natural sound of water exploding out of the ground is nearly drowned out by masses snapping the shutter. I take a couple pictures but keep my eyes on the oddity bursting in front of me.

Water continues to explode from the ground with incredible force and grace. After a few minutes, the water's stream begins to weaken and retreat back into the ground. The hole in the earth continues to steam, releasing its remaining energy before cooling off until the next show.

"That was it?"

"I thought it would be bigger."

"Alright, let's go eat."

It is what it is. There is another one in ninety minutes, maybe that one will live up to your unhealthy expectations.

The crowd dissipates and takes off in all directions. I am tired of being surrounded by people and desire to return to the woods where I can have Nature all to myself, back to where the words of others don't reach my ears. I crave the silence and placidity of the wilderness. I stop at the general store to resupply and eat more food before leaving the spoils of society. On the way out of Old Faithful Village, I walk by smaller, bubbling geysers and past clear pools of scalding hot water.

Further down the path, people going the other way gape as I approach and turn to stare as I pass. My overgrown beard and untidy hair draw as many double takes as my worn-out shirt and skinny legs. They catch a whiff of my musk as I walk by while I turn my nose to their own scents and perfumes. One lady even has the mind to shout out when she sees me walking her way.

"Hey, look! A real mountain man!"

She points at me, not that I'm that difficult to discern in a crowd. All her friends goggle as I approach. I force a small smile and murmur, "Thanks," not wanting to linger and be asked the same predictable and uninspired questions. Old Faithful Village is behind me, and this mountain man needs to be back in the woods, away from all of these tourists and asphalt aided sightseers.

There is a rain cloud spitting above, and the crowds head back to the village to take cover. I have no desire to spend any more time in this clusterfuck and continue toward the protection of trees and the haven of wilderness. The people disperse as I take out my umbrella. I will leave Yellowstone tomorrow after spending tonight just within the park boundary at Summit Lake. Soon after exiting Yellowstone, I cross the border into Idaho and will be able to cross another state off the list.

Part Four: Idaho & Southern Montana

XVI.

CANS AND CAN'TS

August 24 - Day 117

Smoke fills the sky, and there is no one else around. Anything past ten miles out is lost in a haze of gray. I move through the smoke as if it were my own little bubble of protection. When I move, the bubble moves with me. It is only morning, yet the sun is high in the sky, doing its best to shine through the haze. Fires are burning all through the west, and many portions of trail have been closed down. Not this portion. Not yet, at least.

I stop at a short footbridge where overhanging bushes muffle the light trickle of a stream. When all else is silent in Nature, any moving water can be heard as clearly as a steady cascade. Water breathes life into its surroundings, softening the soil, thickening the air, and livening the landscape. I walk up and down the stream, looking for easy access to the flowing water. The bush is thick, but I find a way through. I dip my bottle into the stream seeping from the ground, allowing it to fill slowly with water and many specks of dirt. I dump it out and try again, collecting considerably fewer floaties on the second try.

I take my shirt off and remove my socks, ankle brace, and shoes, laying them out on the footbridge before sitting down to enjoy the sound of the stream. I have been walking fourteen to eighteen miles a day since leaving Dubois, and my ankle feels okay. Not completely painless but still considerably less painful than before. I am fine to take it slow through this challenging section of trail as I take it easy on my body a while longer. The resupplies are still manageable at this pace, and I enjoy taking longer breaks throughout the day.

My things are spread all over the trail, but I have no worries about other hikers needing to pass on this footbridge. The last thru-hiker I saw was ten days ago, way back in southern Yellowstone. I drink what is left in my other water bottle before refilling it in the stream, not being too careful to avoid the floaties that helplessly infiltrate my bottle. It's just dirt, right? I add just a drop of bleach to this water, confident in my gut's ability to handle the small amount of bleach in addition to the dirt and bug shit that leave their own traces in the water.

It is going to be another long day walking along the border of Idaho and Montana. I am not sure where or when I cut north, but it should be soon. I hope it is soon. This border walk is full of tedious ups and downs as it follows the Divide between the two states. I prefer to not think about how much of this lies ahead and instead look at my map to consider how much water I should pack out. This is the last reliable water source for roughly eighteen miles. I will not be going another eighteen miles today, so this is the last water for the rest of the day. It's still before 10:00 a.m. and shaping up to be another hot day in Idaho. Or Montana. Whichever one I'm sitting in at the moment.

State lines have lost meaning to me since leaving Wyoming. I ponder that perhaps I am not in any state and that the border is a larger stretch than the difference between two neighboring blades of grass. State borders are lines. International borders are lines. The CDT is a line. I sit awhile, letting my sweaty clothes dry in the sun while I down another liter and snack out of my food bag.

I max out my four-liter capacity and add a drop of bleach to each bottle before pulling my socks, ankle brace, and shoes back on. I look at my long sleeve hiking shirt lying on the wooden footbridge. At the beginning of the hike, it was a color somewhere between dark maroon and pale violet. Now, it is almost unrecognizable. The back of the shirt where my pack rests and rubs constantly has lost all coloring. The fabric is thinning in these areas, and holes are rapidly developing, becoming larger by the week. Portions of the shirt not covered by my pack hold much less color than it did in New Mexico. This shirt has spent many hours in the sun, which is out nearly every day on the CDT. The sun sucks the dye right from the fabric. It is an exposed trail, and very little is under the protection of trees or tucked away deep in a rivers' canyon.

I cringe at the thought of what the sun is doing to my exposed skin but stopped carrying sunscreen weeks ago. I pull my shirt on. It's dry and slightly stiff. The smell it holds has long since phased my senses. It also fits a lot looser than it did at the start of the trail, from a combination of being worn every day and me losing weight, along with the majority of my upper body strength. Hiking is a great way to tone the legs, shave a few pounds, and exercise the mind, but a lot of other muscles suffer from lack of use.

I pack up my food and water and strap my backpack shut, which bears its own marks from the scalding sun. I buckle the hip and chest strap before cinching them down, attempting to alleviate the eight pounds of water I just added to its weight. Though the pack has molded to my body, it is still extremely uncomfortable. The frameless design has its advantages but also holds many downfalls. Lucky for me, it is falling apart and needs to be sent in for repair. A new, framed pack is waiting for me in the next town, Lima, Montana. Only a few more gruesome days lugging this sack from my shoulders before I get to switch it out.

I have become quite good at pacing myself with drinking water and not being too reliant on such a valuable resource. One of the things New Mexico taught me was to treat water as a treasure, and the Great Divide Basin did well to remind me of this lesson. Now it is merely part of my nature to go a whole day on four or five liters of water, even if the sun is out the entire time. A few times further south, miles were miscalculated, or a water source's reliability was not confirmed. I have learned this lesson multiple times now and am not about to repeat it this far into my hike.

The trail descends toward the bottom of another mountain. I arrive at a road intersection and look toward the climb that sits before me. I can just make out the top, or at least what appears to be the top, which is nearly camouflaged into the sky. The trail follows the road for a short while before turning off to the right. There is no trail sign or post indicating the turn-off, merely a small stack of rocks visible only to those who are looking for it. My well-trained eyes easily recognize the hidden junction and beaten-down grass. I take it without losing a step.

I walk toward the mountain ahead as the trail dips under the shade of scattered trees and winds past CDT markers barely visible on random tree trunks. I coast up the switchbacks and feel fresh hiking in the shade. The sun takes the energy out of me so quickly, but under the trees, my body feels a lot more capable of climbing mountains. I sweat less and keep a steady pace.

After a couple of miles, I reach the first saddle. Ahead of me, the trail twists and turns away, continuing to climb past the protection of treeline. My morning poop is coming on a little later than normal, so I resolve to take a short break while there is still shade available. I drop my pack and head off-trail to find a cozy spot on the hillside to dig a hole. I return to my pack a little lighter and restore my toiletries to their proper place.

Removing my sit pad, I find a nice stump on which to rest. Since most of my breaks involve sitting in the dirt, I try to take advantage of Nature's benches wherever they present themselves. I am slowly increasing my mileage but still not trying to do more than fifteen miles a day. The more breaks I take, the slower I go and the fewer miles I end up doing or am tempted to do.

Soon, I hear voices coming from down the trail. When I turned off the road, I noticed a group on horseback a short ways behind me. I thought they would stay on the road, but we must be headed in the same direction. Their voices echo up to where I sit as they yell to be heard over the trot of their ride. The voices rise with every switchback while I sit, waiting for them to join me on the saddle.

The first two men come around the corner, atop their horses. They are both leading another horse behind them as a dog scampers between the many legs. He approaches me for a sniff but quickly darts off to dig in the dirt. The horses look to be loaded up with at least fifty pounds each. Saddlebags stuffed to their limit are draped off either side of the horses. Both men hold a beer can in one of their hands.

"Mornin'," I greet them.

"Howdy," the guy in front answers, "where ya comin' from?"

"Camped at Lillian Lake last night. Where y'all headed?"

"We're goin' to Blair Lake tonight. How long have you been hikin' for?"

"Today's day 117."

"Holy shit! Where'd ya start?"

"Mexico."

"Are you serious? Hey," he turns around to the guy behind him, "did you hear that? This guy's been hikin' for 117 days!" He turns back to me, "How much further are you goin'?"

"Canada."

"Damn! You're doin' that Divide Trail then? I've never met someone who was actually doin' it. You gonna hike the whole thing?"

"That's the plan."

"Shit, man, I can't believe it. That's crazy. How far do you hike in a day?"

"Anywhere between fifteen and twenty-five miles."

"Wow, I couldn't do that. I can't even walk five miles."

"You'd be surprised what your body can do," I reply, encouragingly. "Have you ever tried?"

"Well, no. But I prefer to ride horses anyways," he says, taking another swig from his beer. "You smoke weed, right, hippie like you?"

"Yeah, but I'm taking a break from it at the moment." A week-long break actually. My mind was becoming foggy, and I wasn't sure whether I was hiking to smoke, smoking to hike, or just smoking because I had it with me.

"How about a pull of whiskey? We got plenty."

I think about it for a beat. "No, thanks." It is hot, and I only have so much water for the day. What good would one shot do me anyway? "How long are you guys out for?" I ask.

"Just a few nights. We do this every year, load up the horses and visit some lakes. This one's the cooler." He laughs and points to the horse behind him. It shakes with the discomfort of carrying a heavy load. I hear ice swishing around on both sides. "Can't go for a ride without beer! So what d'ya do for food?"

"I get into town every five or six days to resupply. I was just in Island Park yesterday, and I'll reach Lima in a few days."

"That's crazy. What else you got in that backpack?"

"Tent, quilt, sleeping pad, change of clothes." I wish they would talk more about themselves, but clearly, they are far more interested in my journey than their own at this point.

"Good for you man, I can't go into the woods without a horse or two carrying all of my stuff."

"Yeah, I only carry what I need along with a couple luxury items. Are y'all from this area? You've got some beautiful land around here."

"I can't do that, I need my things. Yeah, we're from Idaho Falls. Came up for a few days. So what do you do when you ain't hikin'? Do you work?"

"Yeah, I deliver skis in the winter and also worked at a brewery for the past year to save up money for this hike. This is my vacation."

Eventually, their two other friends catch up to them on the saddle, each of them with another two horses following the one they are riding on. Each horse looks more tired and weighed down than the last. These guys also have a beer in hand, aiding the sun in its dehydration attempts. The guy in front turns around and fills his buddies in on what the backpacker is doing out here. They all acknowledge my absurdity before saying goodbye and pulling their horses down the trail.

The ten horses kick up dust on their way past me, and their riders all offer a nod of the head or tip of the cap as they pass. I raise a hand in salute and smile as their steeds carry them down trail. I pull out my journal and begin to write, wanting to be well out of earshot before I move on. They are only going a few more miles before turning off, and I am in no rush to go anywhere.

I'm the first CDT hiker they ever came across. The trail seemed like a myth to them, marked by tiny plaques and the occasional signpost. Surely no one would ever actually try hiking the whole thing in one go? Their attitude of incredulity was expected, but I did not like how many times the guy said, "I can't." It was a constant stream of negativity flowing from his mouth.

You don't know what you can or can't do until you do something that gets you out of your comfort zone. The guy I was talking to kept repeating these words, and it constantly interrupted the flow of conversation. My desire for decent dialogue is still unfulfilled after this gentleman's constant uncertainty of his capabilities, and my half-hearted attempts to stimulate his helpless soul.

I did my best to encourage him that he could do such a thing but narrow-mindedness bars his ability to think outside of the small bubble in which he lives his life. It is difficult to comprehend that a person can go on living in a tiny circumference from his or her home in small-town Idaho but can't fathom something as simple as walking every day. If you can't even walk five miles, I worry about what else you are incapable of doing. Please contain your negativity and self-doubt or else project that shit elsewhere.

If you go through life telling yourself "I can't" to everything that seems odd or out of the ordinary, you're right. Saying "I can't walk five miles" is an insult to your body and mental capacity. It is weak. Those who can walk thirty miles a day or run a hundred miles in a day did not do so on their first try. Like with anything in life, it takes time and practice.

At least be honest with yourself and me. I much rather the fat guy on a horse drinking a beer say, "That's dumb. What the hell you doin' that for, idiot?" than, "That's crazy. I can't do that." It is selfish and small. Find the balls within yourself to do something you did not think possible before. Give yourself a challenge and work toward it. These uneasy situations will give you the jitters and nervous sweats, but it is a hell of a lot more fun than sitting around, getting old, and wondering where the time went.

Perhaps this way of living blinds me in some aspects of life, but it certainly provides light upon others. I only hope witnesses of the thru-hiker lifestyle are encouraged to reexamine their manner of living and the generally accepted habits of society. Thru-hikers can live happily in the woods and out of a backpack for six months, walking across our beautiful country, while some can't escape into the backcountry without two horses to carry them, a hundred pounds of gear and two coolers packed with beer. They are afraid of doing without their things, afraid of listening to the silence, and afraid of having downtime. Staying constantly busy is a gift and a curse, confusing the lines between appropriate application and meaningless effort.

Hiking 2,500 miles is viewed as an impressive feat to many people, no doubt, but when outsiders look at hikers, they cringe more at what we are able to do without and how long we manage to live without these things.

"Everything fits in there?"

"Is that all you're carrying?"

Well, what else does one need? Entertainment is provided by wildlife and trees, by the rising and setting of the sun. I watch the moon go through its phases and have a front-row seat to the procession of the seasons. Being a witness to time is entertainment enough. Being able to rely on no one and nothing except for myself and my body is rewarding enough.

And of a job or employment? At best, it provides a temporary smile, a fleeting sense of fulfillment, and a larger number in my bank account. When employed, I am provided with days of labor that wear me out so that on my days off I have no remaining energy to exert, consoling to take a few laps on the ski hill, before collapsing onto the couch in a heap before noon. On the worst days, I give up my time for a bit of change, then go home feeling worthless to sell myself so cheaply to a cause I care so little for. On the uninspired days, I perform an act of attentiveness, deep down wishing I was hiking trail or sitting beneath the shade of a tree listening to the birds sing. The money that employment offers provides just enough motivation to fake it for a shift. All the while, I think about what I'm going to do after work to make the day worthwhile.

Time is but a tool we measure our lives by. It should be evaluated by experiences and growth, not by the days of the week or the turn of a new year. The stiffness of a monotonous life ensures a stale existence. The unwillingness to put yourself into uncomfortable situations means you will always live within your means as an individual, unsatisfactorily content as time continues to spin away.

No matter. At least they have given me something to think about for the rest of the day. And probably tomorrow too. They leave me in their dust, and I sit on the stump a while longer, getting ready for the exposed mountain and ridge that will be the latter half of my day. Full of rest and fresh thought, I take to the trail and get moving. I leave the treeline too soon and pass the junction to Blair Lake where those guys are headed. It should be an empty trail the remainder of the day. Without the protection of trees, the wind picks up. The trail twists back and forth around the mountain, providing a path of gradual approach.

Visibility is still low. It is frustrating at first, wanting to see sweeping views of the Idaho and Montana countryside, but the beauty filling this ten-mile circumference offers plenty to look at. The higher I climb, the heavier and louder the wind blows. With no trees to act as a barrier, my pack catches these gusts like a sail threatening to nudge me off-trail. For some reason, the wind is never at my back, but always blowing from the side or else straight toward my person. I can feel my frustration climbing with the increase of wind, but stopping to appreciate the surroundings reminds me to relax.

A TRAIL'S POPULARITY

If any one of the long-distance hiking trails can afford to become more popular, it is this one. Without the aid of trail markers along the whole path, a worn-in foot track from repetitive steps achieves similar results. As experienced hikers, we do not require blazes every five feet. Instead, we rely on our ability to recognize the footprints of those in front of us. The imprint of a sole in the dirt, mud, snow, or dust is a blaze of its own and one that is easily wiped away. There are not many people hiking these desolate and removed paths, but as long as everyone stays within a day or two of one another, a hiker won't get lost too often.

The roads that coincide with much of the Continental Divide Trail are already well established and not often traveled. These portions of two-lane trail are ready-made to handle an excess of hikers. They also offer the opportunity to hike three- or four-wide while walking down trail, which is a social advantage neither the Appalachian Trail nor the Pacific Crest Trail can boast of having. Most of these roads are deep in the backcountry and rarely used, making them a good place to set up camp for the night. I have witnessed a few hikers cowboy camp right on the road and remain impressed by the ingenuity as they don't leave a trace by stomping out a bit of grass elsewhere.

Designated campsites are not established along most of the trail, and many evenings end with me looking for a decently cleared flat spot, on top of which I leave an imprint of my tent for a night. When departing in the morning, I cover my tracks and attempt to fluff the grass, but the ground leaves unmistakable traces of a creature having rested. It could have been a large bear if not for the geometric impression left by my groundsheet.

Beside the rough wheels of a truck, dirtbike, or ATV, the footsteps of a hiker are quite tame. Various car parts accompany the many aluminum beer cans, glass bottles, and empty cigarette cartons that litter the shoulders. When you are walking road instead of driving it, you notice a lot more about the landscape, including the mess that is left behind by careless engine-assisted travelers, blind to the beauty beyond the road. Our impact, at least on the roads, is secondary to that of the wheeled beast.

The environment is always suffering at the hands of ignorant people, and a hiker's impact is minimal for the time he or she is out on trail. Our consumption of plastic-wrapped goods may increase, but in comparison to other beings running the wheel, our impact is far less. For rides into town, we rely on friendly four-wheeled travelers, restoring trust in humanity while leaving our lives to the pilot of chance.

Between the roads exist established trail or else faint game paths marked by no more than stacked rocks or wooden posts. Often, I follow a path worn in by wandering wildlife in the area. These faint tracks provide an easy route up the side of a mountain or across a ridgeline. In the desert, I follow cow paths. Further north, it is that of the deer, elk, and mountain goat. These are not always the most noticeable routes, but a trained eye picks up on them fairly easily. The repetitive walking of such a game path benefits from the occasional hiker stamping down the patches of grass that remain, making it slightly easier to identify and follow. Lighter than the cow, heavier than the crow.

When walking on trail, I rarely see anyone outside of the occasional thru-hiker or section hiker taking advantage of the desolate wilderness. The busiest parts of trail are through National Parks or the stretch where the CDT coincides with the Colorado Trail. On either side of the Colorado Trail, I was climbing blow-down after blow-down and traversing across bare ridgelines. Coloradans have the outdoors ingrained in their lifestyle, and while it is always great to see people enjoying the great outdoors, trails are noticeably wider, parking lots are overflowing, and traces remain. As a consequence, state systems and trail organizations are left to clean up and repair the more popular paths, while also doing their best to educate the public on how to maintain proper outdoor etiquette. We are merely visitors and will do well to remember this, especially when in such delicate environments.

As long-distance hiking trails continue to increase in popularity, we must treat these routes with increased respect. A plethora of hikers on the AT and PCT will only result in overflow onto other hiking trails stateside. Exposure from the internet, social media, books, and general word of mouth has encouraged more people outdoors and into the woods. The CDT 3,000 miler list has been increasing rapidly every year since 2012 when less than 10 a year could brag of accomplishing the feat. The accommodation of smartphones and accurate GPS systems has resulted in more hikers attempting to thru-hike this trail when just a few years prior, hikers were relying on paper maps and asked to do a little more navigation than merely following a line on a screen.

The quiet and exclusiveness is part of what makes the CDT what it is, but after the Coalition's efforts this summer to mark the whole trail, they are clearly preparing for an increase in use. None of these trails will ever be the same, and more people are appropriately taking advantage of our public lands. At least the tiny towns that line the CDT will have more visitors to their otherwise out-of-the-way locations. Tourism can be a profitable business, and hopefully, hikers can keep a generally good name for themselves while encouraging the leg-aided travel industry.

XVII.

TINY TOWN, USA

August 27 - Day 120

I wake up around 7:00 a.m. and completely deflate my half-flat sleeping pad. I add "check for leaks in pad" to my town to-do list and feel like I've done this all before. As I sit up in my tent and breathe deeply, the petrichor of a recent storm fills my nostrils. It rained on and off through the night, forcing a restless sleep. Whether I was awoken more by the sounds of the storm or from a leaky sleeping pad remains a mystery. I pack up my small collection of belongings, strap my tent and rainfly to the outside of my pack, and pull my rain suit on. The skies are still dark and gloomy, threatening more rain at any moment. Sensing weather patterns and reading cloud formations has become routine at this point. It looks and smells like rain, and I do not think it's going to clear out any time soon.

I am only six miles from the hitch into Lima. A short day into town is often called a "nearo" because it is nearly a zero, but not quite. This gives me another night in the woods and almost a full day in town to rest and get things done. I also prefer to get a motel room as early in the day as possible to get my money's worth and ensure I get one. These small towns usually only have one motel, so rooms are limited, half of which tend to be occupied by locals staying long-term.

Last I checked (five days ago), this is the day that holds the greatest chance of rain. I want a short day, so I can get into town early and avoid walking in the rain too much, if at all. Last night when I stepped out of my tent to brush my teeth and stash my food bag, hail began to fall for a while. I retreated to my tent and listened to it patter against the rainfly, drowning out moans of nearby cattle.

I am walking by 7:30 and return to the dirt road that will lead me out of these hills. Last night, cows were scattered all around the road, but there is no sign of them now. All too soon, it begins to rain again. I keep my head down and shove my hands in my pants to keep them warm. I can see my breath, and it feels cold enough for snow. I do not want to stop for any reason, not to eat nor to rest. It is only six miles and should take me two and a half hours. I am walking to stay warm and figure if I stop, then I will get even colder than I already am.

A few miles from the hitch, I leave the hills and come out to an open landscape. The rain stops almost immediately as I turn to look at where I just came from. The clouds sit low, resting atop the trees. They dissipate at treeline and tease into nothingness. Ahead of me stretches a long, wide valley, and I see tiny cars moving along the interstate sitting far below. The road I stand on gradually winds down to meet up with it. The sun occasionally peeks out as I pull off my rain suit and unstrap my tent and rainfly. I hold them out in succession to dry as I raise my arms high in an attempt to catch the light breeze created by my stride. Every time I feel raindrops, I walk a little faster, desperate to get into town before the rain leaves the mountains for the valley.

It is easy to see why they call it Big Sky Country and also why the distances between town stops are increasing. There is simply nothing out here. To the north, I see for many miles. Mountains casually drift down into the valley where a large field takes over for the next twenty miles. On the other side of this field, mountains rise out of the ground and look taller than the ones I just descended from. The soft, brown grass glows golden under sunlight in the valley. Peaks standing a ways off rise high, splitting the horizon. I will be continuing toward these mountains when I get back on trail. For now, my focus remains on getting into town. Before arriving at the hitching point, I stop to properly pack away my tent and pull my leggings off. These short shorts and toned legs may help me flag down a ride.

I climb the fence beside the interstate, walk over to the two-lane road, and stick out my right thumb. Cars waste no time in blowing by. I doubt whether half of them even bother looking at me, too focused on the road that lay ahead. People travel on interstates and highways to get somewhere as quickly and efficiently as possible. I figured it would be tough to get a ride here, and my suspicions are soon justified. I smile and wave at every car, excited at the prospect of going into town. Every type of vehicle passes by. Most ignore me. Some wave back. Others shrug their shoulders or else put up a hand in plea. The longer I stand here, the more I realize it may take a while to get a ride.

As vehicles approach, I judge the possibility of each one picking me up. Shiny new Audi with a guy in his mid-forties? Doubt it. Tractor-trailer? Long shot. Big truck with a large driver? Fat chance. Truck with an in-bed camper? Guess not. Pretty girl driving a 4Runner? I hope you like hikers. *WOOSH!* Must be playing hard to get. Stickers cover the perimeter of her rear windshield, amongst which I spot a Backcountry goat, a Patagonia logo, and a 26.2 oval. I have hiked plenty of marathons on trail, surely she would recognize the type. Is this not adventurous enough for you? You think these outdoorsy folk would be aware of the CDT and willing to give me a ride or at least have an open enough mind to help a guy with a backpack out.

I take the shades off my glasses so I can look into the eyes of those passing me. Every person that passes is no longer a dark shape behind a windshield. The trouble with hitchhiking on an interstate is that people tend to be more cautious. This road can take you clear to San Diego if you stay on it long enough. How far does the stranger on the side of the road want to go?

Their speed worries me. Everyone is clearly in a hurry to get somewhere, not wanting to stop for even a few minutes unless they can do so at a conveniently placed gas station and store to refuel both their car and themselves at the same time. It takes me four or five days to go a hundred miles, and these people are doing it in an hour and a half. They are missing out on everything between the big green signs, blinded by reflections of what lay ahead. There are so many things you miss when in a metal cage.

After standing around on the side of the road for thirty minutes, I put my jacket on as I start to get chilly. The wind is picking up, blowing the storm clouds closer. If I stand right next to the road, they'll be able to see me better, right? Most of them slide over to the far lane in case I decide to run out in the middle of the road. They speed by mumbling words of revulsion to their empty cabin. Yeah, I talk to myself too. I catch glimpses of them as they rush past, observing faces of wonder and horror. Some drivers drink from soda-filled styrofoam or pick up their phones in an attempt to make it look like they are too busy to acknowledge me. I know you see me, do not act as if this is all part of a normal day for you.

Finally, a white work van slows down and pulls over as it passes. I whoop with joy and jog after it.

"Where ya goin', amigo?" the driver asks across his friend.

"Lima. It's like fifteen miles down the road," I reply.

"Hop in."

"Thank you so much!"

I slide open the side door. One guy is taking up the whole bench seat. He doesn't bother moving over at all, so I throw my backpack in with the paint cans, brushes, and rollers before taking a seat on a cooler sitting just inside the door. A shelving unit in the rear sways as we slowly regain speed and return to the interstate. A few soccer balls roll around in the back while the paint cans on their side do the same.

"What are you doing out here, man?" the driver asks.

"Hiking the CDT."

"The what?"

"The Continental Divide Trail. It's a long-distance hiking trail from Mexico to Canada."

"Where'd you start?"

"The New Mexico/Mexico border."

"You came from Mexico?"

"At the border, yeah. And I'm out of food, so I gotta get to town to resupply."

"You're crazy, man."

I know.

"Yeah, maybe a little."

With every day I get further from the Mexico border, this hike becomes that much crazier. It did not seem as crazy when I was walking through New Mexico, but it has definitely reached another level up here in Montana. Other people calling the hike, or me, crazy, validates similar thoughts I ponder every day. Having done a thru-hike before, I know living such a life is crazy, but it is the kind of crazy that inspires me. It is also a hell of a lot of fun and the most rewarding thing I have ever done in my life.

These people could have had four more bodies in the van, and I'm sure they still would have pulled over to ask where I needed to go. It is always those with less who are willing to give. They give because it is what they know to do. They may not have much, but when they have any more than they need, they are willing to share whenever and wherever the opportunity presents itself. They're on their way to a job site, traveling two hours to get there. I do my best to educate them about the trail and how I ended up here. We talk about the recent World Cup while cars pass us regularly in the left lane. At the gas station, I hop out of the van, triple-check that I have everything, and thank them for the ride.

The town is small. Smaller than small, actually. It's tiny. Lima has a gas station (with convenience store), motel, cafe, steakhouse, high school, mechanic, Tesla charging station, and a post office. It is all right next to the interstate, and they make their way off travelers driving through and those staying a few nights while on a hunting trip. There is no grocery store, but I did send myself a resupply box here, which should be waiting at the post office. I go there first to pick up my two packages, the other one being a new backpack. The post office is a few hundred yards away, and I make it in just before they close for lunch.

"Hello! I should have a couple of packages here for Brian Cornell?"

"You got here just in time, we were close to sending these back," she says, turning around to pull them down from the shelf. They only hold packages for fifteen days before returning them to the sender.

"Well, thank you for not, I would have been very hungry otherwise."

She hands over my boxes, and I pay for shipping on the backpack. I am extremely excited to use this new backpack and hope at least some of my shoulder pain will be relieved. The frameless pack I have been using gets to be a bit uncomfortable when carrying over twenty pounds. With longer resupplies approaching, I want a pack that can handle the weight more appropriately. At this point, I have forgotten what comfort is. I think I removed the word from my vocabulary somewhere in New Mexico.

Speaking of comfort, I should go get a hotel room to rediscover the meaning of this word, even if only for a night. I walk around the corner to the motel sitting right by the interstate. A boarded-up coffee shack sits on the edge of the motel parking lot. There are a few cars parked in front of rooms, and a housekeeping cart sits in front of room three. Only a couple days prior, I was warned by two southbounders to avoid room three due to its rampant bed bugs. Across the interstate stands a tall mountain, its top frosted with a fresh coat of white from last night's storm. It is not even September yet.

I walk into the motel's office and wait to be helped. On the wall behind me hangs a poster of the CDT, mapped out from end to end. Gazing upon the map, I estimate myself two-thirds of the way through my hike. A small chunk remains ahead of me and tears well up in my eyes. I can't help but smile at having made it this far while flashes of miles past fill my mind.

I ring the bell on the counter, and an older lady makes her way out of the door off the side of the office. Bonnie is extremely welcoming and tells me they have already had the pleasure of hosting a slew of thru-hikers this season. I ask for a room for one night, hoping number three isn't available or ready.

"You can have room eleven, but the door is currently off the frame. Rudy'll put that back in place whenever he returns. He had to go get new hinges for it. Is that okay?" Bonnie asks as she writes down my information before swiping my card.

Fine by me.

"That's fine. There should be a postcard here for me too? My brother came through a few weeks ago and said he left one for me at the front desk."

She turns to a stack of papers beside the desk and begins to shuffle through them.

"Your trail name 'Knots'?"

"That's me!"

Bonnie hands me the postcard along with a key to the temporarily doorless room. The postcard holds kind words from Nacho, Tarzan, and Woodpecker (another hiker I met briefly in Colorado). My brother and I exchange texts or video chat whenever we happen to be in town at the same time, but shoutouts in the logbooks or messages like this do well to lift my spirits.

I make my way across the parking lot to the wide-open room number eleven. As a hiker, you learn to take life as it comes and to only concern yourself with what you have control over. You also develop a certain basic trust toward people, especially in small towns such as this. I drop my stuff onto the one made bed, grab my wallet, phone, and key and head across the street to Jan's Cafe for lunch. It's the only place open to eat on a Monday.

At this point during the hike, I am still vegetarian, and the only non-meat options on the menu for lunch are grilled cheese, potato chips, salad, fries, and pie. I order the first four items to eat there and take a slice of pie to go. Choices are limited as a vegetarian on trail, even more so in these tiny towns that have a limited view of the outside world and the slightest concern of their health. I would have had better options eating at the gas station than at this cafe.

The great thing about these small side-of-the-interstate towns is that everything is only a short walk away. There may not be a grocery store, but at least everything I need is within a quarter-mile. Motel, cafe, convenience store, and post office: the Lima town square.

I head back to the motel and see a few hikers milling about in front of their rooms. When I get closer, I recognize Braveheart. He just got back from PCT Days in Cascade Locks, Oregon, having taken a week off to take multiple hitches and bus rides to meet up with hiker friends and enjoy the festivities. Last time I saw him was at the church in Pinedale, Wyoming. Today, he's getting a ride north to start hiking again.

FourStar is here too, who I last saw in Yellowstone. He is departing today as well and starts talking me through the re-route he's taking directly out of Lima. Wildfires have led to trail closures further north of the next town stop, but he has decided to walk on roads straight out of Lima to avoid these upcoming obstacles. Researching the fire re-routes is on my town to-do list, so I take a picture of his plans for reference. It is good to have options.

After wishing these two happy trails, I retreat to my still doorless room and throw myself on the bed. The skies outside are growing darker, but there has been no rain in town yet. "Watch a soccer game" is on my town to-do list as well, so I prop up my legs and get comfortable to watch a Premier League match on my phone. During the game, Rudy returns to put the door back on its hinges. He fixes it in a jiffy and leaves me to my privacy.

When the match ends, I start the process of switching backpacks. I dump all my things on the bed and hang up my rain suit to dry. I am slightly worried about how everything is going to fit into this new bag. It has more compartments and will surely feel foreign on my back for the first week or two. Movements that have become habit with my old pack are thrown out the window as I began to fill the various pockets, attempting to create some sort of sensible organization along the way.

While I organize my things and go through my resupply, the sky continues to darken outside. The storm clouds finally reach town, and it starts to rain. I watch from behind the protection of my motel room window, glad to be in town, once again thankful for having gotten a ride here in a timely manner. I settle back down on the bed and start my research on the fire closures and possible routes around them.

XVIII.

SPLIT

September 1 - Day 125

I have heard rumors about this section of trail from Lima to Leadore. Very few rumors I choose to believe due to the fluidity of conditions and a hiker's state of mind or personal fitness, but they may be true this time. For 350 miles, the Continental Divide Trail corresponds with the mountains that make up the state line of Idaho and Montana. Upon exiting Yellowstone, the trail winds mindlessly, twisting in every direction, following the state border while offering seemingly endless ups and downs as the trail zig-zags along with the mountainous landscape. Though the views are stunning, this roller coaster terrain is physically demanding.

Even after hiking over 1,800 miles, I am reminded there are no easy days. Each day holds challenges, which is what makes this journey all the more rewarding. With every hill I climb, I peer out across another twenty awaiting my trivial footsteps. Between every hill is a descent as steep as the prior climb, only to be followed by yet another sudden climb up the other side of the saddle. Each up is as repetitive as the down that succeeds it.

The winding of the border makes each footstep slightly less remarkable because my direction is no longer north. It changes with every twist and turn of the Divide. There is no visible footpath, but there are steel posts along the route marking the state border. "I" imprints one side of the post while "M" decorates the other, marking the difference between the two states. Between this space and that space. This dirt and that dirt. This line merely follows the natural border created by the smashing of plates and the upward expansion of the earth, consequently forming a succession of hills that decides where water will end up. On top of the next peak, I relieve myself. Walking from one side of the ridge to the other, urine from this one stream will theoretically end up in both oceans. Eventually.

The trail continues to toe the line between Idaho and Montana while the weather begins to toe the line between seasons. The days are cooling off ever so slightly, and the bitter breeze blows stiffer with every gust. On the exposed ridgeline, I catch the wind coming at me from all directions. It pushes against my progress and threatens to brush me off the side of the hill. I walk diagonally into the wind to stay on course. The sound of rushing gusts fills my ears all day long, and I relish any break in the action.

With the ongoing wind and endless ups and downs, I do my best to breathe and forget all else. I tell myself I'm fit because I am, and the hills pass by without a second thought. There is nothing to do except walk, so walk is what I do. Picking up one foot after the other, placing each one down mindfully, I push anything else from my mind. This is all that matters, and though I still doubt my pace efficient enough to reach Canada before the snow falls, I must keep moving to set myself up for the best possible scenario before everything spirals completely out of my control.

It is September 1st, and I have now been on trail four months. The days slowly shorten with every passing sun, and autumn lingers in the evening air as each day fades away. Though there are no trees changing colors, the nights are getting noticeably colder. When I take a break in the shade, I pull on my leggings and jacket. The temperature is dropping ever so slightly, but the sun and my movement ensure warmth. I am not sweating as much these days. The wind dries most of my body while my back remains perma-damp.

My legs are the strongest they have ever been, but climbing and descending hill after hill with twenty pounds on my back under the blaze of an uninhibited sun is going to be tiring no matter what shape I'm in.

Water is less frequent through this section, as well. Occasionally the ridgeline walk dips between valleys but not more than once a day. My water limit is usually maxed out to ensure I make it to the next source, and to the one after that if something goes wrong. I am still using bleach to clean my water or just not treating it at all if the source looks good enough. Cows are returning to the area, and I desire to have a filter back atop my water bottle. When I am not filtering my water, I have to be a bit more picky about which sources I drink from. This is not the best situation to be in as there is already limited water seeping out of the ground's pores.

A few creeks during this section are marked to be at least a half-mile, downhill, off-trail. So far today, I have skipped a couple of these sources but now must venture off-trail until I find a trickle of water. I avoid these when possible, but they are marked for a reason. Sometimes there is no other option. The next source after this one won't be until later this evening, where I will most likely be setting up camp. I reach the saddle and see an arrow, constructed of sticks, pointing off the east side of the hill. I double-check the guide to confirm the reliability and direction of the stream. For these unavoidable inconveniences, I leave my pack on trail and descend with only my water bottles.

I listen intently as I step carefully down the side of the hill. The sound of moving water should be noticeable above the silence held by the rest of the air. After five minutes, the ground begins to soften. Liquid escapes from the floor, sponging the dirt all around. The land is giving life to itself, providing food for the grass that grows from it and the trees that stand tall further down in the valley. I walk beside the wetter grass until I locate a spot where I can fill my bottles. Floaties are minimal, and I elect to not treat the water. It is coming right out of the ground, and there are no cows in sight.

I return to the saddle where I left the Divide. I take a rest here and suck down a few mouthfuls of cold water, resting with my back to the sun. I unroll my food bag and take out a few snacks: granola bar, sunflower seeds, and peanut butter. My body craves salt, fat, and good old-fashioned calories.

This resupply is the one I sent myself all the way back in Rawlins, Wyoming. My cravings have changed since then, and I wish I had not bought so many Little Debbie snack cakes. They have too much sugar and are not good for prolonged sustenance. I spread peanut butter on top of the granola bar and sprinkle it with sunflower seeds, taking my time to eat this trail delicacy before shouldering my pack and continuing north(ish).

***

Late in the afternoon, I come across a couple who are hiking south. We stop in the middle of the trail to chat under the shade of some well-placed trees. They look tired and beaten down from prior miles of the day. It is a relief to see people hiking the same trail as I with a similar grimace on their faces.

"Howdy! I'm Knots," I say, offering a fist in introduction.

"Hey, there! I'm Bandage," he replies, bumping my first in reciprocation.

"Hi! Sprouts," she says, doing the same.

"You southbounders?" I ask them.

"Yeah, but we're just doing a section from Glacier to Yellowstone," replies Bandage.

"That's a good chunk of trail. How long have you been out for?"

"We started on July 20th. We wanted to start earlier but decided to let the snow in Glacier melt out a bit more. Since we're only doing a section, we were able to and didn't mind waiting."

As a section hiker, it can be difficult to find your stride. It takes a couple of weeks to acquire your trail legs, your gear is not always dialed in from the start, and once you finally begin to feel at home in the woods, you have to leave. This couple is doing a larger section, so this far into their hike, they appear comfortable enough. They carry the customary hiker scent and look the part.

"Yeah, not a great year for a southbound thru. Though I have seen more southbounders than I expected to this season," I admit.

"When did you start?" Sprouts asks.

"April 30th," I reply, "Four-month trail-versary!"

"Ay! Congratulations!" they exclaim in unison.

"You're going all the way then?" asks Bandage.

"That's the plan. It's gonna be close with the weather, but I'll keep goin' until the snow comes. You have to reroute around the fires?"

"Yeah, we took the western route through Salmon. There's a gravel access road beside the highway that made for decent walking."

"Okay, that's the route most people are doing from what I've collected. You guys enjoying this roller coaster ride? Fun, huh?"

"Oh, you are gonna love the next section, just you wait," says Sprouts.

"I don't know, the past two days have been pretty difficult. Y'all better tighten your shoelaces!"

We each think the section we just came from is bound to be the worst of the stretch of trail for the other. By speaking this aloud, we manage our expectations of what is now in the past and what we have yet to hike. No matter what difficulties lie ahead on trail, we hold on to the fact that the worst is behind us, and it is bound to get better from here. If you can make it through today, you can make it through tomorrow.

"The next few days are going to be more of the same for you," she affirms.

"Lookin' forward to it! I mean, I've made it this far. Can't be much worse than what I've been hiking the past couple days. At least the views are incredible," I say, raising my arms up and looking around.

"Yeah, it's more of the same quick, consistent, steep ups and downs. Any good places to camp over the next couple miles?" Bandage asks.

"Not really. I've just been cowboy camping at whatever flat point I can find on the ridge. No campsite I can really point out or recommend," I reply.

"No worries, we'll find something. Let me tell you about what's coming up for you..."

Bandage then goes into a detailed description of the next ten miles north of here, basically what they hiked today. Sometimes I will ask a southbounder about a town, water source, or alternate, but generally, I am fine to get on without a step-by-step account of what is forthcoming. This is the only thing I actually dislike about talking to southbounders: all of the information and spoilers they feel the need to share. What the terrain is like, what the next town holds, when to take lunch, where the next water source is, and why I should set up camp at a particular place.

I have made it this far without anyone warning me what is ahead and how to handle it, but thanks for letting me know the trail takes a left, then a right, climbs another hill, descends to a water source, and then follows a road for a while. I would rather discover this on my own without all of your details. My supplies, experience, and the map app have done well to counsel me thus far. Any additional information is unnecessary. Especially in the digital age, pictures, videos, and words can be found detailing most of the trail. This ruins part of the experience for me, and I relish walking into the unknown with nothing except what is on my back and the confidence in my abilities to be able to make it through to the next resupply point no matter what I am confronted with.

"...Well, good luck man, hope you make it to Canada!" he finally concludes.

"Thanks! Enjoy your section, don't have too much fun."

We wave goodbye and take off our separate directions. Southbounders are usually good for a chat. I find the conversation comforting because they understand the inspiration and can appreciate the ups and downs experienced. Part of the joy that comes from a long hike is encountering people from all over the world and from all walks of life. Along this lengthy, narrow path, encounters are inevitable. Most long-term hikers I meet are living the best days of their life and know it too. This couple seemed a little odd but were generally jovial and smiled all the same. In the woods and on the trail, we are all just humans carrying backpacks. Familiarity of the deed grants a certain undeniable connection between walkers of the wilderness.

There is a common joke between hikers that all southbounders are a little off because they are clearly hiking in the wrong direction. On all three of the main long-distance hiking trails in the States, the common direction to hike is from south to north. It is mostly easier this way due to weather patterns and the timing of the seasons.

When thru-hiking the CDT, it is best to start in March or April. This is to avoid intense heat in the New Mexico desert while allowing sufficient time for the snow to melt at higher elevations in Colorado. Arrive too early in Colorado, and the snow can be too deep to efficiently walk through. Arrive too late, and you may be hiking into monsoon season, facing more difficulties further north still. Generally, spring allows enough heat and sunshine to melt most of the snow to make hiking more manageable, but when hiking the Continental Divide, you are bound to walk on snow at some point. It is merely another obstacle you have to prepare for and adapt to. Leaving Colorado, the hot summer days can be tough through the Great Divide Basin in Wyoming, but the last remaining challenge is making it to Canada before the snow falls in Montana.

If you go the other way, the challenges are reversed. Start south from Glacier in summer when the snowpack has melted down enough, avoid monsoon season in northern Colorado, make it through the higher elevations of southern Colorado and northern New Mexico before the snow falls, and reach Mexico before the desert's already scarce water sources completely dry up in the late autumn/early winter.

Southbounders are also known to provide faulty information, and therefore I take everything they share with a grain of salt. Whether this is on purpose or by accident is still unknown. The act of fear-mongering is common, but this can be boiled down to our differing directions and reaching certain milestones at varying points in our hikes. Just because I had a difficult time doing big miles through New Mexico does not mean southbounders will. By the time they reach that state, they will be nearly 2,000 miles into their hike, and the flat finish will be easy walking for them. Conversely, they claim Glacier National Park and the Bob Marshall Wilderness to be the best sections of trail, having seen those sections first, yet they have not experienced the Wind River Range, Great Divide Basin, Gila Wilderness, or San Juan National Forest. Every hike is different, and they are especially different when hiked in opposite directions in contrasting seasons.

The trail carries on exactly how Bandage described it to me. I attempt to put it from my mind, but every change in the path comes as predicted. The route continues to follow the state border, and there is no significant trail. A slight pattern of beaten-down grass can be made out when looking across the ridgeline but is hardly noticeable below the feet.

The border is pretty obvious. It is the tallest collection of peaks strung together by slightly lower saddles stretched out to the north. The next four peaks stand in front of me, taunting my progress and waiting patiently to haunt my next couple of hours. I consider walking around the side of the hills to avoid the tedious process of gaining elevation just to lose it, but the brace wrapped around my right ankle pleads otherwise. Uneven terrain caused this injury, and I have no desire to test out my weakened tendons just yet.

It has been nearly three weeks since getting checked out by the doctor in Dubois. There are still occasionally sharp jolts of pain shooting through my ankle, and I take two ibuprofen with each meal. I have not rolled my ankle in three weeks, nor have I hiked over eighteen miles since leaving Dubois, averaging just twelve miles a day. With terrain like this, I probably would not be doing much over twenty miles a day, even if I were completely healthy. The distances between towns are increasing the further north I go. This resupply is one hundred and three miles, so I am forced to gradually up my mileage to avoid carrying eight days' worth of food. The ankle is feeling strong, so I might as well test it out a bit and embrace a little pain every day.

When walking up hills, I move my ankle as little as possible. Preferring to keep it stiff, I use my hamstring muscles to pull my leg up. Part of me wishes I still had a trekking pole to aid in the steep descents, but one of my reasons for disposing of it in Lima is because it was detrimental to my balance and technique. Descending with one hiking pole would have me twisting and stepping awkwardly down hills when all I need to do is walk a little slower and control my movement.

Climbing to another peak, I see the ridgeline zig-zagging away from me. It is raining further off. Ten miles away, I'd say. By the time I get there, it will probably have moved on. The clouds' path is noticeable, and my distance offers a favorable perspective of the weather. I catch a few sprinkles on my head and shoulders but continue on, just behind the tail of the storm.

The ridgeline carries me through the afternoon, and by evening, I begin the descent into the next valley. Another string of mountains waits patiently in the distance. Behind me, there is nothing I haven't seen before, but the sunlight shines kindly on my future. Soft brown mountains look as if they were dripped out of sand. Rising gradually out of shallow land, they slope upwards to a point as its opposite mirrors the approach. The Continental Divide is still obvious, and while I am excited for my legs to carry me north, the border walk remains a challenge. At least the views are pleasant, and the route is scarcely traveled. The reward is ever-present, and after another full day of walking, I am pleased with my progress.

XIX.

SEPTEMBER SHIVER

September 5 - Day 129

Yesterday evening I did not eat any dinner. I went to bed with the shivers and an aching stomach, which resulted in night sweats and restless sleep. I must have some sort of bug because today I feel like crap, and my shit has been getting wetter with every discharge. The four miles to the road this morning were not easy. Stuck in a haze of sickness and determination, I barely remember any of it.

I must have drunk some questionable water over the past week. I have been tiring of putting bleach in my water to kill the bacteria and, as of late, have not been so careful about which sources I take from. Without a proper filter, most of the water I have been ingesting recently has been untreated. This far into my hike, I thought my stomach could handle such distresses, but the combination of ibuprofen and bleach may have temporarily wrecked my insides.

Today, all I want to do is lay down, rest, and drink clean water. I have my tent set up in the yard at Leadore Inn, and I just got off the phone with my mother. I asked her to mail out another layer as it is getting much colder at night. If I want to finish this trail, I am going to need some cold-weather gear in order to camp comfortably. Laying down in the grass at the Inn, I begin to look at the upcoming trail and possible alternate routes. This next section is where the trail closures begin.

One of the things the west has that the east does not are wildfires. While thru-hiking the PCT or CDT, it is extremely difficult to make it border to border without encountering any fire closures along the way. Down in New Mexico, I was just ahead of the late spring fires that forced reroutes, but I was unable to hike through all of the San Juan National Forest because of the blaze that ravaged the wilderness just outside the Silverton area.

Wildfires are bound to be an issue at some point during a western thru-hike, but you never know when or where they are going to pop up. This can make things a little complicated, but the resources are out there for hikers to get around these fires safely and efficiently. I find the spontaneous planning to be enjoyable and a welcome challenge.

There are two sections of trail currently closed between Leadore, Idaho and Anaconda, Montana. The CDT Coalition provides fire closure information on its website along with a recommended detour. North of Bannock Pass, I would have to hike around one fire and rejoin the CDT, only to be rerouted around another closure just after it. This suggested roundabout route adds a handful of miles while circumnavigating the fires yet seems an inefficient way to avoid these closures. I decide to improvise and make my own route.

I look at maps, suggested detours, reroutes, roads, and information shared by other hikers. The official CDT loops around to the west, continuing along the Idaho and Montana border before turning east/northeast toward Anaconda. I check the Continental Divide Bike Route, which follows a road south out of Anaconda, taking to the Pioneer Mountain Scenic Byway until it gets to Grant, Montana, where it continues east. Another option I consider briefly is to simply hitch a ride to Anaconda, around all the fire closures. Some hikers do this when faced with the prospect of a lengthy fire reroute, but I decide to walk the distance.

A thru-hike is defined as connected footsteps from border to border. If I hitchhike around this chunk, I would just have to return later to complete my thru-hike, if and when they reopen the closed sections of trail. I do not mind roadwalks, and they have valuable lessons to teach any hiker. The CDT thus far has taught me to appreciate any marked trail, and walking the roads just makes you enjoy backcountry trail that much more. Fire closures are part of the challenge of completing a thru-hike in the west, and variety is always welcome.

After much research, I decide to walk the road north from Bannock Pass (where I hitched into Leadore) to Grant. From here, I will proceed to follow the Scenic Byway up to Wise River, where I will then turn west and make my way to Anaconda along another road. The Continental Divide Bike Route follows the road from Grant to Wise River, through the Pioneer Mountains, so the scenery must be nice enough to look at. I take a little more time to research what towns are along the way and what kind of roads I can expect to walk on. Once I hike north out of Bannock Pass, I will be done with the Idaho/Montana border. This means I am hiking through Montana the rest of the way to Canada!

With flatter miles and no roller-coaster border walk, I tell myself I can do twenty miles a day. There is a state park, hot springs resort, and a few smaller towns along the way. I still do not have a water filter at the moment, so I resolve to carry a full supply of clean water from town to town to avoid drinking cow and automobile infested run-off from the side of the road. Bleach will be used in case of emergency only. It will be hot and exposed, walking the road, and I'll have to be cognizant of my water intake.

Doing this roadwalk will get me around both fire closures and should put me in a much better position to finish this hike before the snow falls. This route is more direct, and although less scenic, it holds its own challenges. Also, with this sickness I am currently fighting, it is reassuring to be on a road in case I start feeling a lot worse and require medical assistance.

There has already been plenty of roadwalking on the CDT, and I am prepared to embrace a change of pace from the roller-coaster terrain still haunting my memory and legs. The only thing worrying me is that I will be walking at least twenty miles a day on an unforgiving surface. I presume most of the roads to be paved, and this will certainly be a test for my ankle and shins, which have been known to tighten up after a few days of pounding pavement. With the monotony of roadwalks, it is going to be primarily a mental challenge. Master the mind, master the body.

***

With a 131-mile roadwalk looming ahead, I decide to take a zero the next day. My body deserves a much-needed rest before taking to roads for the next week while I continue to fight this bug that has me shuffling to the bathroom every couple of hours. I spend the remainder of my time in Leadore drinking water, sitting on the porch talking to Tom, owner of Leadore Inn, waving at cars driving past on the highway, and eating very little. I call a couple of friends to catch up on the recent weeks and share stories. They provide encouragement and boost my spirits.

The town of Leadore does not have a grocery store. Other than a restaurant and gas station, there is the post office and two motels. People going through Leadore are usually doing just that, and those who do stop in are either lost, tired, hikers, or hunters. There are no other hikers here while I am, reinforcing my aloneness and solitude. I'm reminded that while on trail, there is likely to be nobody within a day or two of me. Though I may not actually have the whole trail to myself, it certainly seems that way.

I have begun to see pictures of those ahead of me reaching the northern terminus and completing their thru-hike. The bubble of hikers that are on pace to finish are currently doing so. My brother should be reaching the border within the next few days to complete his Triple Crown of American long-distance hiking trails. Many of the hikers who are finishing this trail are also completing their Triple. The CDT is usually the trail that most save to hike last, as it is claimed to be the most difficult and challenging.

The Appalachian Trail and the Pacific Crest Trail hold their own unique challenges, but this beast of a hike is seen as the ultimate goal and climax of long-distance hiking trails in the States. The AT is well-traveled and has the luxury of towns every few days, with three-walled shelters every five to twenty miles. The PCT is extremely scenic and has many "flat" miles between the mountain passes that are crossed every day or so. I put "flat" in quotations because the trail is never really flat, they are just not up or down miles which is a great relief for the legs.

The CDT, on the other foot, is roughly marked, remote, exposed, and can be a lot longer. There are no man-made shelters, and most towns are either walked right through (forcing more road miles) or else a decent hitch away. Weather patterns are unpredictable. and help is usually pretty far away. There are some out here hiking the CDT as their first long trail, but for most, it is their second or third thru. Experience plays a huge role in one's comfort and confidence on trail. The basics of a long hike remain the same, the only things that change are mileage, towns, terrain, and personal mindset.

Triple Crowners have completed an impressive goal with over 7,500 miles of total trail hiked and around a year of cumulated time living out of a backpack and sleeping in the woods. This is no ordinary accomplishment, and when I first went hiking on the Appalachian Trail, I remember being astounded when someone said they hiked another long trail. I was even more impressed when I met someone on trail who was finishing up their Triple or had already hiked the three trails once and was now hiking them all a second or third time. After completing my first thru-hike, I figured it would only be a matter of time until I got my Triple Crown.

With every passing winter and approaching spring, I consider getting back on trail. Before this spring, the timing was never right, and I never had enough money saved up to make it work. I was also worried about having a place to live for the winter that I stuck around all summer just to lock down my room for another year. The process of finding a decent place to live is difficult in a small mountain town, and the situation I had was worth holding on to for a couple of years.

Time between hikes convinced me I would hike all three long-distance hiking trails at some point in my life. The four months I spent on the Appalachian Trail were easily some of my best months ever. I enjoyed some great months since, but the AT changed me and set a life course that would be difficult to discover or justify any other way. It was only a matter of time until I got back out on trail. Given I complete this one, I will just have the PCT remaining to achieve the Triple, working my way east to west while saving the easiest and most scenic walk for last.

Backpacker Law states, "You either do one long trail or all three." If you hike one long-distance trail and have a terrible time, you are likely never to try again. Or you're simply satisfied with the one and have no desire to go back to living out of a backpack for multiple months. Once you hike the second trail, you're only one away from the Triple Crown, so why stop now? You know what a thru-hike entails and how difficult it can be, yet for some reason you have chosen to put yourself through this pain once again, because you clearly enjoy it, believing it to be of some benefit. Once you do two, you might as well do the third, and why shouldn't you?

I have no idea how many miles remain in this hike, but I will have a better idea after completing this roadwalk and returning to the red line north of Anaconda. I do know my route is shorter than the suggested detour to the west, and this smart-cut will allow me to bust out a good chunk of mileage over the next week. With the main bubble of hikers currently in northern Montana, I remain seriously behind schedule, flirting with the impending doom of snowfall. Instead of struggling to do twenty miles going up and down the border while hiking around fire closures, I will be able to do at least this every day while walking north, for the most part. The route I am taking goes through a couple of small towns that are sure to have restaurants and a decent enough resupply. Since the towns are within a day or two of one another, I won't have to carry much food on top of the four liters of water I'll be lugging around.

A zero in Leadore should allow my body some rest while this questionable bug continues to wreak havoc on my system. It could be my body breaking down after being pushed to its limits for the last four months or perhaps some combination of bad water and bleach on top of the ibuprofen I was ingesting daily for my ankle pain over the last few weeks. There is no doctor in Leadore, but I make sure to drink a lot of fluids and buy a few more packets of Imodium. All I can do is relax, hydrate, and hope for the best. This is my first time getting sick from bad water, so I guess my good fortune has run out. It was only a matter of time.

XX.

PERKS OF A ROADWALK

September 8 - Day 132

One good thing about climbing hills on the road is that the grade is usually calm enough to where I barely notice the incline. The only downside to this particular climb is that I can't really see the top of it. I just passed the town of Polaris, Montana, and am glad I did not ask Mother to send a package there. The post office is only open from 9:30-11:30 a.m. and closed on Sundays. Instead, I asked her to send a package with warm clothes to Wise River, Montana, which I should reach in a couple of days. I was also hoping there would be a diner or restaurant of sorts in Polaris, but it is more of a ghost town than anything.

Elkhorn Hot Springs Resort is a handful of miles away, and this is my goal for tonight. More days of hiking should end with a soak in a pool of hot water. The last hot spring I was able to take advantage of on this trail was all the way back in New Mexico's Gila Wilderness. That one was more mud than water, and I could barely sit down in it. I am expecting a little more out of this established resort.

Mile markers beside the road taunt my pace as cars coasting by do the same. The constant reminder there are faster ways to travel ridicules my decision, not only to walk the road but the entire country. The taunting is nothing more than that as I would rather be the crazy one on the side of the road, slightly hunched over from the weight of a backpack than the lazy commander of a four-wheeled machine made for getting between destinations as quickly as possible. Purposeful walking is my preferred way to see the country and to experience the land, without the filter of looking through a window at blurred trees beyond.

The shoulder alongside the road is minimal, and I remain as close to the edge as possible, counting the beer cans that litter the roadside. I walk on the left side of the road, facing traffic because it is what you're supposed to do. This way, drivers see me, and I see them. If they are distracted and don't actually see me, I can react by getting the hell out of the way. Most oncoming cars slide over while others barely slow down, ignorant of the empty lane on the other side of the yellow line. Regardless, I always wave a hand of thanks. Thanks for not running me over. Shadows behind windshields gently lift their hand off the top of the wheel in salute as rubber rolls them swiftly out of sight.

My feet are sore, nearly to the point of numbness, but just enough feeling remains for me to feel every little piece of gravel protruding from the otherwise flat asphalt. I look down at my feet and try to walk on the white strip of paint. This thin white line is softer than the surrounding black and noticeably cooler as well. There is no noise out here except for the passing cars that reassure me this road leads somewhere. I hear them approach and continue to listen as they drive off to a soft rumble that carries through the valley. A vehicle is approaching from my rear. The dull roar subsides to a hum as it pulls to a stop beside me.

"Hey, man. You need a ride?" The driver has sunglasses on and is speaking to me over the AC blasting inside his shiny 4Runner. His buddy in the passenger seat leans forward to get a better look at me.

"No, thanks. I'm good," I reply.

"You sure? There are some hot springs a few miles up the road that we're headed to. We can give you a ride if you want."

"Yeah, I know. That's where I'm headed."

"Okay. You need anything? Water? Food?"

You got any beer in that cooler? "No, thanks, I'm all set."

"Alright. You sure you don't want a ride?"

"Yeah, I'm sure. I gotta walk."

They look confused. Who would walk this road voluntarily?

"If you say so. Be safe."

"Thanks."

I hold up two fingers as he pushes the gas pedal down to carry his load the remainder of the hill. The incredulous looks I get when turning down rides is part of the fun of walking road miles. I appreciate the humanity to offer the dirty, forlorn backpacker on the side of the road a ride, and I applaud the offer of water and food. I turn you down, not out of spite or hostility. There is no bitterness flowing through my veins, no feeling of vanity. I walk because I want to. Sure, if I accepted a ride to the hot springs, I would be there sooner, but what would I do with the rest of my day? This walking is an activity that fills my time with pleasant progress and meaningful movement.

Maybe I will see you at the hot springs, in which case I will gladly accept your company and share a few stories. But, all the more likely, you will probably have already gone by the time I get there, moved on to other things seen as important or time-sensitive. My pace is my own, and it has carried me thus far. I walk for the solitude, and I walk for something to do. I rather spend the next four hours walking before soaking in the hot springs for an hour than soak for an hour and spend the next four looking at the inside of my tent. I am in no rush to get anywhere, only happy to be where I am.

My stomach is mumbling words of hunger, but the snacks in my food bag are not appetizing at the moment. Eating dried food out of a bag for six months is almost as difficult as walking trail for six months. The afternoon is beginning to look more like evening, and my cravings build with every passing house that lines the road. I try not to get my hopes up, but a feeble fantasy of being invited into one of these houses to be fed occupies my imagination for a couple of miles. After seeing a small billboard advertising the Grasshopper Inn Bar and Grill, my fantasy is interrupted to entertain images of a mouthwatering burger, a tall pile of french fries, and a healthy plethora of vegetables. On the left in a mile and a half.

The perfect burger manifests itself in my mind and begins to tantalizingly rotate on a golden platter. A thick, rotund patty, topped with french fries, bleeds out to be soaked up by a soft, yet durable bun. Perfectly sliced lettuce, tomato, and onion combine with the brown palette to complete the colorful combination of meat, vegetables, and bread. Lettuce and onion crunch with every bite, providing complementary texture and flavor. One large slice of tomato moistens the bun above, holding its head high despite drowning in its processed brother's blood. Ketchup and grease escape the confines of the circular construction, dripping onto the plate. Two buns, although slightly different in shape, team up to offer a hand in maintaining a culinary creation that would crumble without their aid.

I do not normally entertain food cravings when walking as it is a cause for distraction and unrealistic expectations, but it certainly is a good motivator when walking road miles. I walk because I enjoy being out in the woods. Sacrifices, such as eating out of a food bag, have to be made in order to accomplish this. Granted, I am not out in the woods. I am on the road and will be for the next few days as well. The sacrifice of walking in the woods has been made in lieu of this roadwalk so I can finish my hike at an appropriate time.

Further up the hill, I finally reach the Grasshopper Inn. I know how long it takes me to walk a mile and a half, and the estimate on their billboard seems to be a bit off according to my finely calibrated legs. It is the hottest part of the day, and I have not taken a break in a while. The Grasshopper Inn Bar and Grill is just the haven I've been holding out for.

I walk the long driveway and stop outside on the deck to remove my shirt and put my jacket on. It may not make much of a difference, but it's something. My jacket holds considerably less smell than my hiking shirt, which permanently reeks of B.O. from gallons of sweat ingrained in the fabric. I am incapable of smelling my own stink as I have become one with the stench. The odor of a thru-hiker is a musk to wear proudly, capable of turning noses down a supermarket aisle or clearing out the corner of a bookstore.

I walk dazedly into the restaurant as my eyes adjust to the lack of light inside the building. There is no one else seated, but an older waitress emerges to greet me.

"Have a seat anywhere you like," she tells me.

I take to the nearest booth, which also has an electrical outlet underneath the table. She brings me a menu as I ease off my backpack, ordering a Coke and ice water. I dig out my charger and plug in, chugging the water the moment she drops it off at my table. It is too late for lunch and too early for dinner, but as a hiker, any time is a good time for a meal. The sickness I picked up before Leadore is still affecting me, and my stool has yet to harden, flowing out of me with worrying ease. All I can do right now is drink clean water in an attempt to flush my system and take advantage of the facilities while I'm here.

Today is day 132, and my body feels worn down. My brother and I hiked the Appalachian Trail in 131 days, so this is now the longest I have ever been out on trail. Every day after this is an accomplishment of its own as a testament to my desire to be out here as long as possible. In order to do this, my body must be properly tuned physically, mentally, and nutritionally. I have been increasing my mileage on this roadwalk while chasing a deadline that seems improbable. This lingering bug has disturbed my system, and I must replenish what has been lost. There is no doubt I have lost weight during this hike, and my body is certainly lacking proper sustenance at the moment.

Though I was hungry for the past couple of hours, I refused to eat anything more than a bar as I held out hope for a roadside restaurant. For the last few days, I have been craving a burger, and today is the day I break my nine-month trial of vegetarianism. My cravings for meat cannot be ignored any more than my body's internal protests against another damn plate of french fries. The famous Idahoan potatoes are a cut above any other spud tossed into the fryer, but I feel they are not offering much diet- or nutrition-wise when I am walking twenty miles a day. The hiker hunger has reached a new level.

I order a burger with all of the fixings with fries as a side and salad as an appetizer: one last salad before I rear my canines back into action. I ask the waitress to bring a pitcher of water as I sit there with two empty glasses in front of me. Soon, she brings my food out, and I dig in quickly to avoid the suspense of my first bite of meat in nine months. Nine months is not that long, but I am still proud of myself for having made it this far, especially amidst a thru-hike. I do not consider how my return to meat is bound to affect my already screwy digestive system, but I hope the vitamins and nutrition offered in exchange will offset the damage.

The burger is delicious. It comes out just as I imagined it, save for the golden platter. Even with nine months of anticipation built up, it meets all my expectations. It is more fulfilling than a stack of sliced spuds and more flavorful than a bunch of salad, though I devour these as well. I check the map to see how much further it is to Elkhorn Hot Springs and confirm with the waitress for conversation's sake. Waitress says 3 miles, but the map says 3.6. I am glad I checked, or that last half mile would have been a cranky one. It looks to be all uphill. With a full stomach weighing me down, I pull myself back onto the road and continue along the windy asphalt trail up into the mountains.

***

A few cars pass me. With the heat of the afternoon gone and the cool evening present, the miles are easy, even with a slight incline. Often, I find a shallow slope all the more walkable as the ground rises beneath my feet. Elevation gain of an inch with every step accumulates quickly. Without any other cars stopping to ask if I need a ride, I soon reach the entrance to Elkhorn Hot Springs Resort. The resort has cabins, campsites, hotel rooms, and a restaurant all on their premises and is much more than I was expecting. The smell of geothermal activity pulls me along the gravel road until I reach the pools where a heavy steam exudes into the air.

I briefly consider soaking a while before carrying on to stealth camp a little further down the road, but at this point, it is nearly dark, and I have already walked twenty-six miles to get here. Marathon mileage is plenty for one day, and I would much rather enjoy my time soaking than trying to find a decent place to sleep in the dark. I pay for camping which includes use of the hot springs and walk even further down the road to find an open site.

My shins are starting to feel tight again, and I have acquired a limp since entering the resort boundaries. The shin tightness is slightly worrying because this is something I was dealing with before taking a week off in Durango. No doubt this has developed over the last few days, from walking road, but at least I am able to soak in the hot springs tonight. From previous experiences with these enchanting pools of water, I am confident a long soak will make everything better.

The nearest open campsite is just on the other side of the trees, about a quarter-mile from the pool. Voices carry from the pools to my declared campsite for the night. I hope these disappear with the appearance of the stars. A bear pole hangs between two trees, and I set my backpack against a log perfect for sitting. Right away, I begin the well-rehearsed process of unpacking my belongings and erecting my tent. The flattest spot is in the gravel pull-through, but I much prefer to sleep on the grass, on softer ground, and removed from the possibility of an intruding vehicle in the late hours of the night.

Before I know it, my tent is up, my sleeping pad is inflated, and my quilt warmly awaits my purposeful intrusion into its depths. Still full from my elaborate dinner, I put my food bag into my odorless sack and walk it away from the tent, stashing it in the bushes fifty yards away. I have had no problems with bears thus far and am confident in the noise of this place to deter unwanted visitors. I am also counting on other campers to have a greater and smellier selection of food, having perhaps just cooked weenies over the fire or steaks on the grill.

I write in my journal and look through the map while waiting for darkness to fall. The hollering soon subsides, and I walk gingerly back down to the pool area, rinsing off in the shower before taking to the tubs. The smaller, shallower pool is a spicy 106 degrees while the larger, deeper pool is around

92 degrees. I walk to the larger pool and put my hand in the water. It's more warm than hot. I walk down into the smaller enclosure and feel the heat rising off the surface of the water. This one is just right.

Hot water from a large pipe falls from head height and slams down into the pool. I sit on the bottom of the tub with my head barely above the water and my back comfortably against the wall. The water crashing into the tub is loud and creates cover for conversation just as jets do in a jacuzzi. A man sits on the other side of the stairs, and we have to yell over the raucous while repeating ourselves too many times.

The man I am talking to is an Elkhorn resident. Locals occasionally treat tourists and visitors alike with a certain coldness, holding resentment toward those who have exposed the secrets of their home. Guests of the resort will doubtlessly boast of their trip to others, encouraging more visitors to dip their toes in the Elkhorn pools. Sharing your waters with strangers can be a warm welcome and a great place to share a drink or conversation, but some places must be protected for their own good.

As a thru-hiker, I usually get the benefit of the doubt of being a good person and adventurous enough to be applauded for my efforts. I walked to these hot springs from Mexico, dammit, if anyone needs a soak, it's me. Other visitors to these tubs merely drive a few hours to reach this destination and linger for days at a time. The local and I share good conversation about the outdoors, skiing, and life. Darkness shrouds our identities, while the waters heal our bodies. He asks me what my favorite part has been so far. Today.

I begin to stretch in the pool. More stars expose themselves by the minute, and I have lost track of time since stepping into the water. The pools close at eleven, and I may just stay here until they ask me to leave. After stretching for a while, I move over to where the water is crashing into the pool and raise my right leg out of the water to meet it. The hot water massages my shin and soothes the pain. I switch legs back and forth to quiet my tense muscles. I turn my back to the water and let it pummel my neck and shoulders.

If there was a hot spring like this at the end of every day, I would never have any injuries. The mystical hot spring waters of Elkhorn have temporarily cured my body and mind. Eventually, I climb out of the pool with pruned fingers and hydrated skin. I am not sure what time it is. What I do know is that it's bedtime. I walk back to my campsite, moving more comfortably than I have the entire day.

Part Five: Northern Montana

XXI.

RETURN TO TRAIL

September 13 - Day 136

The first yellow leaves I saw were on the second day of my roadwalk. At Bannack State Park, a couple of trees stood with yellowing leaves, unsure of themselves amidst a crowd of green. There were only two of them making this premature progress, standing out plainly amongst the evergreens. Autumn is my favorite season. The slow, gentle process of trees shedding their leaves, as they change colors with the passing days, is hardly matched by anything else. Trees and shrubs undergo this annual exercise of purging their greenery before embracing the frigid days of winter, willingly allowing decorations to fade like a rainbow overtaken by another storm.

At first, I was thrilled to see trees with bright yellow leaves, but my excitement quickly drifted into subtle anxiety as further proof of the advancement of time before me. Further along the road, patches of gold spatter the distant hills. Among the constant green of the Montana countryside, fading leaves and shrubs conspicuously leap out of the landscape. These sights are pleasing to my eyes but indicate the actualization of autumn's approach. There are officially ten days left of summer, the heat of which has suspiciously disappeared over the past couple weeks.

My frustration with the roadwalk stretch increased as the asphalt pounding leg carried on well past its date of enjoyment. Miles were becoming just that, and the scenery around me would be constantly wounded by the large hunks of metal thundering down the black strip dissecting the land. I am extremely pleased with how well my body held up, and my mind is just as glad to be back on dirt. Upon opening the map app and loading the northern Montana section, I made a beautiful yet terrifying discovery: I have less than five hundred miles to go. With an average of twenty to twenty-five miles a day, I will be done in a few weeks!

I certainly picked up my pace over the past week, but I also cut out a decent chunk of twisting, climbing, turning, and indirect miles, that would have no doubt been wonderful to stumble through while waiting for winter's hands to tear me away from a brief autumn. It was a difficult roadwalk section, but completing it has put me in a great position to reach the border by the first week of October. Time is not completely on my side, but there is still a chance.

***

My return to the woods is as brilliant as a sunrise yet as subtle as a shift in the wind. I follow the road out of Anaconda which gradually rises and twists through the mountain until it encounters the CDT at an intersection. I turn to the north and see a large marker nailed to the nearest tree. I run to the tree and wrap my arms around it, planting a kiss on the steel plate before bouncing down the dirt track.

After seven and a half days of trudging down the open road, I have at last returned to trail. CDT markers are never all that common on this trail, yet there are plenty throughout this section. The last trail marker I saw was before the hitch into Leadore, and their return provides a calming sensation of being back where I belong. I seem to have forgotten how quiet it is in the woods with no cars around threatening to run me over and a forgiving dirt trail to soften my footsteps. There is no noise besides my backpack bouncing behind me, and the sound of each breath carefully taking in fresh, undisturbed mountain air.

The air is much cleaner, and the ground is void of window tossed trash. Post-roadwalk, I am reminded of Nature's necessity and how much better it is to walk a trail than the shoulder of a road. This distinction was never questioned, but the relief is undeniable. A rhythm is reestablished on trail, much different than the daily routine of the roadwalk. Spirits are high as I am engulfed by the warm embrace of a gentle forest. The calmness of the trail provides comfort as I find solace amidst the silence. My thoughts speak louder, and their intentions are clear without the impending *woosh* of a passing car.

On the road, I felt that passersby were missing out on all of the beauty beside their paved path of choice. Back in the woods, I feel I can't walk slow enough to enjoy all that contributes to the Montana forest. Trees of all sizes are in different stages of growth. Some are so skinny I can wrap my arms easily around them, while others stand so thick with age that my wingspan is not sufficient enough to accomplish the same gesture. The sky is partially blocked from branches dangling overhead, providing shade and shelter. With my eyes constantly looking from side to side and up and down, I remember to take a deep breath and absorb all I can. My energy is back where it needs to be.

The simplicity of trail overwhelms me as I adjust to walking the faint dirt path that lies ahead. The road had me following the destruction of man across the once beautiful lands of Montana, with large green road signs standing intrusively beside the road and mile markers counting off every mile. A dirt path in the woods and the occasional trail marker are all I need. Sure, these paths and trails are used for recreation, for those who want to talk a walk for the day or spend a few nights in the backcountry, but it is mind-blowing to think that any combination of trails, along with the occasional road, can get you pretty much anywhere in the country.

Any travel after this hike will be viewed differently, thanks to the unique perspective granted to me from this adventure. There really is no hurry to get anywhere. If you are so worried about getting somewhere, then you are there and not here where you should be. Meanwhile, the mind finds consideration of the past and future an easy excuse to explore the trenches of memory or desire. It travels to different dimensions, timelines, and alternate universes of never-ending possibility while the body remains in the present. Force the mind to live in harmony with the body. Be where your body is and appreciate the beauty of the moment.

Afternoon skies bring dark clouds as I prepare for the possibility of precipitation. Next minute, the sky breaks open with a crack of thunder, dropping heavy rain and small bits of hail. I could keep walking, but there is a large tree nearby, so I elect to sit this one out, retreating to the protection of the many-layered branches overhead. I sit against the trunk with my pack beside me as I enjoy a snack and watch the rainfall. Listening to raindrops splattering the ground, my thoughts drift off toward less important things.

Most of the day has been spent with my jacket on. In Anaconda I checked the weather, and it is not going to be above sixty degrees for the next few days. It is officially summer until the 22nd, but I am starting my goodbyes now as I embrace every ray of sunshine I am fortunate enough to feel. The rain subsides just as a chill begins to creep up my spine. I depart the tree's radial protection and walk with my hands stuffed into my pockets.

A few sprinkles spit out of the sky while the higher elevations wear a crown of fog. Low hanging clouds tickle the treetops, as I walk underneath hoping for nothing more to become of the forbidding grayness. As evening approaches precipitation returns. I wear my rain jacket for the rest of the day while my feet enjoy a thorough washing from falling rain and standing puddles. Everything is covered in water as I step carefully over slippery rocks and logs. I have not seen the sun for a few hours and do not expect it to make a return before I finish hiking for the day. Even without the sun, the forest glows in the rain. Leaves glisten with the sweat of the sky while portions of the trail turn into a mirror of the treetops.

Later in the evening, I start looking for a decent campsite. Sometimes I end up walking another two miles before such a campsite presents itself while other times I find one around the next turn of trail. I crest over another hilltop and notice a flat spot between young trees just to the east. I remain on trail and check my map to consider the terrain ahead and the possibility of another flat spot further on.

There is a long downhill coming up, and I have gone roughly 24.5 miles for the day. Good enough. The downs tend to be as difficult, if not more so, than the ups, and my legs are tired. I can feel the sun setting as the sky continues to darken above me. The chill in my bones brought on by the storm lingers. With a break in the rain, I figure it the perfect time to set up camp.

I head toward the flattest part, and barely have time to set my pack down before I am rushing to take out my trowel and wet wipes. I had a few close calls with diarrhea while walking the road but managed to avoid making a mess in my shorts even with minimal warning. Again, I have little time to dig a hole before pulling my shorts down and letting whatever I had earlier that day make a quick escape. It is definitely still loose, but when I stand up to check what came out, I lose my appetite for any dinner I was thinking of force-feeding myself. A small load of mucus sits in the shallow cat hole. It looks like a sick moose's snot rocket or a chunk of phlegm hacked up by a bronchitis carrying bear smoking a pack a day.

My body is still fighting off the virus I picked up over a week ago. Hopefully, this is the worst of it as I am not sure what other disgusting items my body can expunge from its system. I gave up on taking anti-diarrheal medicine on the roadwalk, and although I bought some Pepto-Bismol tabs in Wise River, I have declined to take them over the last few days. Whatever I have is not going to be treated with a couple of pink chewables. I do not have my Sawyer filter yet but will be getting it next town stop.

The small dropper bottle of bleach remains in my backpack. I continue to drink the cleanest water I can find, hoping it to be void of any more nastiness while allowing my body to rebuild its defenses against bad bacteria and welcoming back the good. The inside of my gut needs time to recover after being wiped clean from bleach and ibuprofen. This is all theory as to what is happening inside my body. No knowledge is confirmed, and any suspicions are from instinct or else researched from the internet. My best guess is that it's Giardia, but it really doesn't matter what they call it. It sucks, and I hope my system can reset itself soon.

I could be completely wrong about this self-diagnosis and in dire need of antibiotics and medical attention. I am sure any doctor or nurse would have a better, more educated response to my symptoms, but I would not know out of personal stubbornness and general distance from any medical practice. My self-prescribed medicine is to walk outside for twelve hours a day, chew my food completely before swallowing, and drink lots of water. So far, it is working, the only symptom being persistent diarrhea.

This body has been through a lot over the past 136 days. I have been rewarded by sipping the sweet nectar perspiring from Nature's overgrown pits and offered the most beautiful views atop her perfectly shaped contours. Conversely, I have been beaten down and punished by the painful track of the Continental Divide corridor, with road and trail alike combining over distance and time to inflict beyond normal wear and tear on a body formerly thought fit for such an undertaking, leaving nothing for me to drink save for a few murky puddles remaining in deep troughs bearing any number of diseases within its volume.

My appetite for dinner is gone, not that I had much of an appetite, to begin with. Trail food is becoming less appealing by the day. Whenever I force myself to eat another granola bar, I only see it as necessary nutrients and calories. Most of the foods I eat are smothered in peanut butter: one of the few food items I have yet to tire of. Recently, I've taken to eating peanut butter right off of my spoon, licking it like a viscous popsicle that never melts.

For the past few days, I have not worn my ankle brace for hours at a time in an attempt to restrengthen my ligaments and tendons. The fact that I made it through the roadwalk without any major issues, narrowly dodging shin-splitting shin splints and the worst of this sickness, is encouraging. My body is hanging on at the moment, and I am sure it could use a few days rest before I take to the Bob Marshall Wilderness and Glacier National Park. I only need two to three weeks of decent weather to finish the last four hundred miles. This final push will be all I have left though I feel there is so much more to experience and see. All things, whether great, terrible, or mediocre, are temporary and must come to an end.

Before any of that happens, I will be taking a couple of zeros in the city of Missoula. A friend from Mammoth is enjoying an adventure in Glacier with some of her hometown friends, and they have an extra ticket for a concert because one of their numbers bailed on the trip. I'll arrive at MacDonald Pass the day before the concert and could not have timed this any better if I tried. I also want to check out Missoula because it could be a place I move to after finishing my hike. I have heard some good things about this city, and it checks a few of the boxes on my list of post-hike preferences.

A couple of days off in Missoula should provide me with adequate rest and a resurgence of energy. Though I enjoy my solitude in Nature, certain emotions and energies are void in the wilderness. Trees and plants create a buzz, but the buzz of people out and about the town or wiggling around on the dancefloor is a completely different energy. I have been starved of this human energy along with the interactions and feelings that accompany being surrounded by thousands of other persons. Perhaps I should be more cognizant of my days and how important each one is in the timeline of finishing this hike before October, but the break is much needed for morale and will surely by my last set of zeros before I push into the Bob Marshall Wilderness and continue through Glacier National Park.

Before getting back on trail, I will pick up one last mail drop of supplies from the Elliston Post Office. My mother has been an excellent trail angel throughout this hike, sending whatever I ask for along with a few snacks and always including a note of love and inspiration. This final delivery has my Sawyer filter, sleeping bag liner, and passport. If I am going to make it to the border, I better be prepared for whatever I may encounter along the way. Having already picked up an additional layer and beanie in Wise River, these few additional supplies should get me to the border safely and warmly (enough).

It is no longer raining, but it has gotten dark far earlier than it should due to the clouds still low in the sky, blocking the sun's radiance. I walk my food bag well away from my tent, take off my wet socks, and dread pulling them back on in the morning. Alas, that is tomorrow's problem. I crawl into my quilt, hoping the morning sun rises in a favorable location, blessing me with its first light through the thicket of trees around me. I close my eyes and tell my body it is strong and healthy. This may not be the truest of statements, but positive reinforcement has been known to be effective when there is no other option.

The last three days have held cloudy skies. Rain has fallen on two of those days, and the temperature has failed to get above sixty. The goal is within reach, and I keep telling myself I will make it to the border because there is nothing else I can tell myself at the moment. In town, I will visit thrift stores to find another pair of leggings to help keep my bottom half warm as the daily high gradually decreases. It is going to be a cold few weeks, but this is the final stretch, and no injury, knock, or bug will slow me down.

XXII.

A SHORT AUTUMN

September 25 - Day 148

Days are growing noticeably shorter, while the landscape's change in color is apparent everywhere. Yellows litter the land and provide motivation to keep walking. Oranges fill the hillside and shiver in the wind. Though there is less sunlight as autumn carries me further toward October, I am walking some of my biggest miles of the whole trip.

The colder temperatures keep me moving while the finish and impending snowfall rush me along. I barely take any breaks during the day apart from a morning shit break and an afternoon lunch rest. When I stop moving, I get cold and have to put on more layers to stay warm. It is much easier to keep walking and not have to worry about layering up just to de-layer soon after.

There are very few people out here, save for some locals out for a hunt or one last camping trip before settling in for the long winter. From my short conversations with these outdoor enthusiasts, I am encouraged to keep walking until the snow starts. One guy camping with a trailer gave me serious eyebrows when I told him I was headed to Canada, which was around three hundred miles away at the time.

Heeding their winter worries upon me, these seasoned Montanans share stories of years past when snow fell during the last week of September. They provide warnings of how the weather can change in an instant, tossing a thick blanket of white over everything in sight overnight. The skies above hold foreboding and warnings of their own. A few days ago, while walking over the tallest peak of the day, I was welcomed with snow flurries floating down from above, being whipped around by the wind.

On the hitch out of Augusta, I met a guy who was going in for a hunting trip. He told me I should be fine until at least the first week of October. Most people are in consensus on this and suggest a similar approach as the season wears on: check the weather, be prepared, and don't be stupid. Easy enough. I have made it this far and believe myself capable of hiking through a little bit of snow but also know I too have limits. As a thru-hiker, your head can inflate a large degree, but one must keep their ego in check when preparing for a difficult or dangerous section. While I check the weather for Augusta and East Glacier Village, no weatherman is providing a report from deep in the Bob Marshall Wilderness.

I slept beside the river last night, falling asleep to the calm gurgle of water constantly in motion. I rise early and need my headlamp to see in the dark. Packing up camp takes no more than fifteen minutes as I collect my food bag that is hanging from a dead tree on the way out of camp. Yesterday, I resupplied in Augusta and packed out eight days' worth of food. It should only take me six days to walk to East Glacier Village, but there is a chance of rain in the forecast, and I want to be prepared for whatever weather may fall from the sky.

I am carrying so much food that it won't all fit in my food bag. A plastic bag filled with ramen, tortillas, baby carrots, and apples dangles off the back of my pack. The last couple of weeks, I have been hiking a few miles before eating breakfast, wanting to get moving as early as possible and because my hands stay much warmer when I tuck them into my jacket pockets. Despite the higher mileage, I enjoy these cooler days and walk comfortably without sweating for most of the day.

The sky lightens just after 7:00, but the sun does not rise above the mountains until at least 8:00. When I am walking before the sunlight touches me, I feel accomplished and ready for the day. The sky offers a show that takes time to reach its climax. The dark blue fades as the yellow ball of light lifts ever higher. The clouds hold no intention of moving and can't help but shine in appreciation of the day.

When the sun comes up, I stop to shed a few layers and squeeze some water into my cold-soak jar, moistening my breakfast mush of oats, peanuts, sunflower seeds, and chocolate morsels. With this eight-day supply of food, I am eating as much as possible. I could probably make it last ten days if I have to, so I am not worried about rationing my food just yet. Carrying so much food is a pain. This is the heaviest my pack has been since I got it in Lima, but I rather buy more than enough food than not enough, especially when hiking such a removed stretch of trail through the Bob Marshall Wilderness.

The clouds do not look to be clearing up anytime soon, and I maintain my pace as the trail continues to follow the river. The rumble of small rapids rolls gently to fill the valley with a calming tune. My feet step softly on the dry dirt, providing no warning to the deer I sneak up on. They use the trail as a path of least resistance and snack on the remaining greenery lining the corridor. I stop walking to allow them space to eat and so I can watch them graze. Whether due to my loud hiking habits or their acute awareness, I have not seen much wildlife in Montana besides the occasional deer and a handful of elk. The deer eventually notice me and bounce away down the trail, retreating from our meeting place.

I rock-hop a stream and make my way into a large, cleared-out campsite area complete with logs perfect for sitting. I take a break and eat a second breakfast, desperate to lighten my load as quickly as possible. A little further along, I come to another stream that is wider than the last and looks to be at least knee-deep on the far side.

I take my shoes and socks off, pull my leggings up above my knees, and elect to walk across barefoot. I absolutely refuse to soak my footwear so early in the morning. With no sun visible, it will take days to dry these shoes out, and a brief spell of cold feet is better than suffering through a whole day of wet and cold feet. I stuff my socks into my shoes and hold them tightly in my right hand. Making sure everything is secure and packed up, I step into the stream.

After taking three careful steps, my feet lose feeling in the frigid water. The first few steps are well-measured, but once I can't feel my feet anymore, I move quicker, oblivious to the rocks I step on. I now wish I took more time to walk up and down the stream to at least find a crossing that is a little shallower. Already halfway across, I am past the point of no return. The water is up to my knees in the middle, and at the far side, it reaches my thighs. The deeper crossings were enjoyable in New Mexico, and even in Wyoming, but these Montana crossings will test my determination. Or stupidity.

I reach the opposite bank and step out of the water, leggings slightly wet and feet completely numb. I sit down in the grass, set down my dry footwear, and pull my quilt out of my backpack. After pulling my socks back on, I stuff my feet into the quilt, shaking my legs while attempting to wiggle my toes. I reach into the bag and use my hands to massage my feet, hoping to bring them back to life and pass some body heat along to them. After five minutes, I realize this approach is futile. The only way my feet are going to regain feeling is if I start walking. I return my quilt to my bag and step my numb feet into dry shoes.

As I walk, my feet begin to regain feeling. My toes loosen with every step and color, I presume, returns to the tips of my tiniest extremities. Whoever said walking can't solve everything is only partially wrong, although walking did get me into this situation in the first place. Walking gets the blood flowing. Walking keeps the body active, and any physical activity allows the mind peace. When in doubt, just walk. When cold, walk. If you are already walking and still cold, walk faster.

The first day of autumn was a few days ago, and the evidence of summer's demise lies everywhere. With each day, the colorway of Nature dwindles as color fades from the mountainsides. Bright, golden yellows darken toward a mustard brown. Glowing, radiant reds subtly turn a mellow maroon. Beside them, tamaracks show evidence of winter's measured approach. These beautiful deciduous conifers radiate golden in the late morning sun and are remarkably noticeable amongst the ponderosa pine.

The evergreens remain just that as their stoicism disguises the changeover of seasons. I walk in the shade, through dense groves of ponderosa and lodgepole pine. They stand proudly throughout the hills and valleys, unconcerned about losing their coat of green or dropping needles. I wait patiently for winter, hoping it will hold off a couple of weeks still. The hills burn in color, autumn quickly becoming a dying season.

The Bob Marshall Wilderness is remote and fulfilling. A six-day stretch in the woods is enough time to get far away from society and removed from the noise. Seeing one or two people a day allows the mind to rest and wander. Enjoying quiet miles, I listen for airplanes and wildlife. A few goats scramble on the steeper parts of the mountain, taking routes only they care to tread, leaving tumbling tallus in their wake. They notice me down below and retreat further up the hill.

Every step north takes me further into winter. I walk into the future as the cold weather settles around me, ensuring each day is darker than the last. Every step up trail is colder and more omniscient than the one before. With less sun comes less green. Less daylight leads to less life in the vegetation as we all anxiously await the first freeze of the season. I walk up each mountain pass to witness the gradient of colors gradually dimming. The hills are silent. The trees don't speak.

I ask for more days like today. Clouds in the sky provide protection and offer company. They glide across the sky as I try to keep up with them far below. The trail follows the Chinese Wall, a hauntingly stunning escarpment, for the latter part of the day, before dropping off the far side of the range.

My body feels great, and my mind is at peace. I continue talking to the trees, and they do well not to interrupt. I thank the clouds for their company and the dirt for its acceptance. I thank the sun for its hope, the trees for their oxygen, and my body for functioning. I know this weather cannot last forever, but I will continue to enjoy it while it is here.

A perfect autumn day it has been. The forecast shows a chance of rain, though a chance of rain also means there is a chance it will not rain. With the progression of the seasons, not only are these chances increasing in probability, they flirt with the possibility of bringing snow to higher elevations when rain is predicted in town. I am not worried as I still have plenty of food, but time is not on my side. Once the snow starts falling, it will be difficult for me to continue.

Today has been the best day. I figure I have less than two hundred miles left and must savor every step, gust of wind, ray of sunlight, raindrop, and snowflake along the way. If the weather remains on my side, I will be done in ten days or less. I have put myself in this precarious position of pushing the limits of the hiking season because of my pace throughout the past 2,300 miles.

Hiking on the far edge of autumn can be dangerous but also rewarding. The reward is more time in the wilderness, fewer people on trail, and more days to witness the changeover of seasons. Life is slowly leaving the trees, and the coats of deer darken with the landscape around them.

XXIII.

SNOW PROBLEM

September 27 - Day 150

I set up camp just in time. Soon after pitching my tent and finding a tree to throw my bear line, it begins to rain. It was cloudy, chilly, and gray all day, but at least the precipitation managed to hold off until the evening's offering. This campsite has a lot of flat, open space under its resident pine trees. I sit under the largest one to eat my dinner until it gets too cold to be outside any longer. At a certain point, my hands move slower with every scoop of ramen that disappears inside my mouth.

I hang my bulging food bag and retreat to my quilt to cozy up for the night. I spend a good twenty minutes massaging my calves, one of which has developed a serious knot. *This body is strong*, I tell myself, as I attempt to knead the pain away.

Before going to bed, I register my progress, thoughts, worries, and wonderings in my journal.

I sleep soundly through the night. At some point, the spattering rain becomes a soft snowfall. Whereas raindrops fall from the sky with ferocity and purpose, leaving the maddening cries of clouds behind, snow drifts gently down from above, suspiciously subtle. The puzzle pieces of a white quilt gradually assemble on the ground floor, coating leaves, branches, and trail. I was worried this was going to happen, but it is just a light covering and not nearly enough to inhibit me from moving forward.

The possibility of taking a trail zero remains in my mind for only a moment. If snow fell during the night, more could be on the way. By day's end, I want to be as close to Glacier National Park as possible. I pack up my frost-covered tent, taking a few breaks to blow hot air into my hands. It will be difficult to dry this out today, but I'll take any opportunity I can to do so. For now, I let it hang off the outside of my pack. My food bag is a couple hundred yards north on trail, so I pack this up on the way out.

I begin the day wearing my leggings and jacket, rain suit remaining stuffed away in my backpack. Not even a full inch fell overnight, but the ground is still mostly covered. Blades of grass poke through, trying to keep their heads above water.

The trail is slightly overgrown, and the added weight of snow has pulled trail-side vegetation over the path. With every step, my legs brush up against the lingering branches causing them to drop snow onto my legs and feet.

Within minutes my feet are soaked. I don't want to stop so soon after getting started for the day, but I need to at least put my rain suit on. There aren't any good places to stop and sit without being in the snow, and there are no trees around with a dry berth beneath them. I brush some snow off a log beside the trail and take a rest.

I pull my rain suit and a couple of plastic bags out of my backpack. I wrap each foot in a bag before shoving them into my shoes. Hopefully, this will keep my socks dry. I struggle to keep everything out of the snow and end up accepting the fact that things are going to get wet as I dig into my food bag for a bite to eat. I don't want to stop again anytime soon, so I might as well stock up on snacks.

Any noise I make is instantly soaked up by the buffer of snow cast across the ground. The trail is lined with bushes waiting to be knocked of their burden. The CDT is not always the most well-cared-for trail, so while this may be due to the snow weighing the vegetation down, I believe this stretch of trail to be naturally overgrown and made worse by last night's snowfall. My shoes are soon soaked through, and my feet begin to feel wet. The plastic bags have already ripped from my aggressive morning pace.

The first part of the day is spent in this nature, with thoroughly soaked feet and a battle between being amazed at the weather yet frustrated by it. I make it through by telling myself what I always tell myself when faced with an obstacle on trail: *just keep moving.* I am not going to be any warmer or drier by standing still, so I may as well keep walking. It has snowed, and there is a chance it will snow again tomorrow.

I do the math in my head and calculate how many days it will take me to reach Canada if I walk twenty miles a day. Twenty-five a day. Thirty a day. Hell, at this point, I could just walk sixteen hours a day, sleep for eight and manage forty miles a day. The previous conversations I had with locals sit uncomfortably in the back of my head as their warnings echo in my ears.

They all claim the first snow to usually fall within the first week of October, if not by the first day. Before even stepping foot on trail, I read on every blog and website that you should absolutely be done by October 1st. Some have tested this deadline and succeeded while others have paid the consequences for hiking out of season. This overnight inch was only a dusting, but the forecast holds a greater chance of precipitation over the next few days. All I can do today is walk and hope the serious stuff holds off a while longer.

My legs are in great shape and feel much better after last night's self-massage. My backpack is finally beginning to feel more manageable as my eight-day supply of food has dwindled to a three- or four-day supply. I'm glad to have bought extra food. Even though only one inch fell last night, it could have easily been one foot. I am grateful for the warning from Mother Nature and re-do the math in my head.

Roughly 125 miles to go. Five days of hiking twenty-five miles a day. Four and a half days if I can manage thirty miles a day. I'll restock in East Glacier with a week's worth of food and get right back out on the trail. I can do this. I have to do this. Every town visited, every mile hiked, every bag of ramen eaten, and every night camped for the past five months has been in preparation for this last resupply. *You can do this, Knots.*

There's only one way I can end this hike, and that is by connecting my footsteps to the Canadian border. I spent six months preparing for this hike, saving up money, researching, buying gear, mentally preparing, and telling others what I would be doing for the summer. I said I would be hiking the Continental Divide Trail, so this is the only thing I can do. There is no point in trying to do something if you do not believe you can. Something can be said for pushing your limits and wanting to see what you can handle as an individual, but without belief, it is not going to amount to anything.

Halfway through this hike, and especially after my injury in northern Wyoming, I began to doubt my ability to reach Canada in time. While the goal never seemed within reach, I was always getting closer. I constantly made progress and kept moving, knowing that if I wanted to have any chance of finishing this thru-hike, I would just have to keep walking. There have been many times along trail where I did not think I would make it, partly because I believed it and partly because I didn't want to be disappointed if I did have to end my hike early due to injury, weather, or lack of fulfillment. I said I was going to hike the CDT so, dammit, I'm going to do it. All of it.

Later in the afternoon, I take a break by the river where there is no snow. The trail has cleared up for the last ten miles, out from under the cover of thick trees and a dense forest floor. I hang up my tent and rainfly on dead trees lying beside the trail. The sun is not out, but there is a light breeze, and this is the best spot I have come across all day to dry things. Also, I am hungry. I take my shoes and socks off before digging out some food. I have been walking pretty much non-stop since this morning and could use a rest before finishing out the day.

Thirty minutes later and the breeze has done a good enough job of drying my wet gear. A bag of Wheat Thins lies open beside my pack. The fallen trees listen to my concerns as I speak them aloud so I can hear them myself.

"It snowed last night," I begin, "and there's more snow in the forecast. East Glacier Village to the border is roughly ninety miles. Get to town, fill your food bag, check the forecast, and get back on trail. You gotta finish this damn thing, Knots. Five days. You can do this. What else are you gonna do? Go home with ninety miles left? You can't bail without going through Glacier. You've made it this far, just finish the hike, dummy."

Things are made much more real when you say them out loud. Here, the trees are my witness. I pack up my gear and pull on my footwear. The sun exposes itself from behind the clouds for a full minute before being taken over again by gray curtains.

Toward the end of the day and closer to Glacier National Park, I hear a group approaching from the south. They are speaking loudly over the noise of jingling reigns and stomping dirt. When they get close enough, I step off the side of the trail, allowing their steeds plenty of room to pass.

"Good evening!" I say as they approach.

"Howdy," says the man in front, tipping his hat. He is followed by a woman and another man, all around their fifties.

"How's the ride?" I ask.

"Great!" answers the lady in the middle. "I feel kind of lazy not walking myself, but it's so much easier this way! Where are you goin'?"

"Canada."

"Oh! You're almost there!"

"I know!"

"Enjoy the rest of your hike!"

"Thanks! Enjoy your ride."

Their horses carry them past me and further down trail. I wait until their shouts are a couple of turns away before I start walking north again.

I *am* almost to Canada. You're right, lady. I responded in excitement, but what is there to be excited about? After they walk away, I repeat the words to myself and wallow in the sound.

"I'm almost to Canada. I am *almost* to Canada."

I am excited to see Glacier National Park, but ending this hike is not something I look forward to. There are many reasons I have been taking my time during this hike, and while I know I can't hike the CDT for much longer, I dread my return to society. I dread returning to the other world when I know the feeling of walking trail for twelve hours a day. Out here, I am a witness to the day, distracted by no more than the path of the sun across the sky, where my only responsibilities are eating food, drinking water, and walking miles. I listen, breathe, think, and appreciate. My time on this trail is nothing more than moments just like now, where I can be in total awe of where I am, and how lucky I am to be here.

One of the great tragedies in life is that all things must come to an end. Yellow leaves covering the dirt remind me of this, as does the white fluffy stuff settled upon the landscape and the burn areas scattered about the mountains and valleys. Everything is temporary: fear, pain, solitude, love, life, warmth, cold, and everything between. Even thru-hikes have to come to an end at some point. The trail is only so long, and the sun won't shine forever. This does not mean we shouldn't pursue these miracles of the world, but a mere reminder to enjoy them in the moment and to not take any of it for granted.

XXIV.

HOSTEL FEVER

September 29 - Day 152

Less than a hundred miles from the border and only ten from East Glacier Village. More snow fell last night. After sleeping on the wet and cold ground, I rise early, pulling on wet socks before stepping into wet shoes. Today, I will be getting into the village to stay in a hostel, check the forecast, and reassess my situation. I begin the day with intentions of getting out of the woods as quickly as possible.

There is more chance of a wintry mix today on top of the precipitation that fell the last two nights. I eat breakfast while walking and will not stop moving until I get to town. Clouds hang low, and I have yet to see any of the spectacular Glacier peaks I feel dominating the horizon around me. Hills rise to the sky but are covered by the masks of weather.

Hopes of walking well-maintained trail into the village are quickly diminished when I come to a clearing where tall grasses and limp branches disguise the dirt path hiding underneath. I curse to the sky and trudge through, sacrificing my rain pants along the way. The once small rips over my knees catch overhanging limbs and extend to the waistband. Holes in the shin area tear to the lower hem. The shreds dangle from my waist, but I keep them on to offer my legs some protection from the onslaught of sharp branches and snow-soaked grasses. These rain pants have lasted me nearly the whole hike. Less than a hundred miles to go, and they fall apart.

I arrive at a trail junction and read the signpost to see how far I am from East Glacier Village. Still 7.7 miles out. My phone remains tucked away in my backpack. I shouldn't need it with these National Park signposts providing all the directions. I hear some people approaching and am surprised to see a few trail runners come around the corner.

"Morning! Great day for a run, huh?" I ask them.

"It could be worse!" the tall blonde guy replies excitedly.

They all wear leggings and a light jacket. They all look cold.

"I didn't think I'd see anyone else out here today," I say. "Where ya' headed?"

"We're goin' up to Firebrand Pass for an accumulation check. You on the CDT?" the shorter girl asks.

"That's right! Have any of you stayed at the hostels in East Glacier?"

They all shake their heads.

"No. We live in Whitefish, so we're only an hour and a half away. But go to Serrano's, they have bunk rooms behind the restaurant. They should still be open. Really good food too," the other girl replies.

"Sounds good, I'll check them out. Y'all training for anything?"

"Le Grizz in a couple weeks!" the guy replies.

"Fuck yeah. All of you? Wicked. Well, which way ya headed? We should keep moving, you guys look cold," I say, getting a little cold myself.

"We're goin' that way," the shorter girl answers, pointing up the hill.

"Me too! After you," I say, stepping out of the way, allowing them to take the lead.

They take off and regain their pace. My backpack is the lightest it has been in a week, so I take off at a light jog behind them. I do not want to be out here any longer than I have to today, and jogging helps warm me up. The faster I get through all of this overgrown brush, the sooner I will be able to eat a warm meal and have a hot shower. Their pace motivates me as the girl in the back takes pictures of those in front of her and even turns around to capture a few of me. I must look ridiculous. My rain pants look more like a hula dress, my hands are stuffed in Wheat Thin bags to keep them warm, and I am holding my umbrella above me. We share a few more brief exchanges before the next junction. They turn onto another trail while I continue straight.

This quick, chance encounter brightens my spirits to know I am not the only one out here. These folks came up from Whitefish and want to enjoy this park as much as possible before the snow makes running trail even more difficult. Soon after they turn off, the trail is silent again. I jog some more, but the overgrown trail slows me down as I reach a field that no longer has a visible trail or even a trace of one. Stomping over brush and shrubs, I attempt to ignore the coldness starting to develop in my toes.

It really is beautiful out here. The mountainsides remain green and yellow underneath fresh snow. Leaves have been getting darker every day, but not all of them have departed their resident branches. The trail opens up in a few areas, looking out into valleys and toward mountains that aren't completely visible. I do not expect to see the sun today. Any chance of me taking a trail zero was immediately tossed out of the backpack when the snow began to fall. It has gotten worse over the past two days, and I need to get into town so I can be out of this weather. Too many days of hiking trail such as this leads to clothes never drying and toes turning black.

***

September 30 - Day 153

Koozie and I are going for a day hike, no-packing twelve miles of trail from Two Medicine to East Glacier Village. I met this fellow thru-hiker the day prior, and we both stayed at the backpacker's hostel behind Serrano's. The hostel and restaurant are both closing today, so we walk across town to pay for a bed at Brownie's Hostel before ditching our bags and stuffing our pockets with snacks.

With only nuts and candy bars weighing us down, we hitch a ride north to the Two Medicine road crossing and begin our hike south back to East Glacier Village. It rained in town yesterday, which became snow at some point during the night. We want to see how the trail looks at elevation.

As the trail climbs up and past the treeline, visibility gradually reduces until the horizon blends with the sky to create an uninterrupted canvas of gray. At the higher elevations, snow is knee-deep, and snowdrifts on the ridge are thigh- to waist-deep. A few inches in town quickly becomes a few feet on the pass, and we carry on with feet that aren't wet, just cold and covered in snow.

On the ridge, with no signposts, trail, or cairns in sight, we navigate using our phone's GPS. Though we have not gotten a proper look at the mountain from town, it being helplessly covered in clouds, we know some mountains in Glacier possess sheer faces. With such poor visibility and no way to discern flat ground from sloped, we can easily walk to our doom.

"You see anything up there?" I say to Koozie, who stands twenty yards ahead of me on the ridge, looking into the whiteout aimlessly.

"Not really," he calls back.

"I don't think the trail goes up the ridge any farther," I say, phone in hand. "We're off the red line."

"You think it turns off back there?"

"Well, it definitely doesn't go up the ridge."

We struggle to find the correct route in fear of misstepping and sliding off the side of the mountain. Thankfully, the GPS is accurate enough that we are able to navigate blindly by staring at our phone and carefully placing each step. With no packs, we need to make it up and over this one pass in a timely manner.

The day is only growing older, so we keep moving out of necessity and fear, a combination that may sound dangerous though it certainly pays off more often than not. We make it back into town just as evening approaches and the temperature begins to drop. My beard is frosted over, and my toes are beginning to lose feeling. We thought the twelve miles would take us four hours, at the most. It turned into a six-hour adventure with travel reduced to three miles in two hours over the crest, in the deepest and most difficult section of trail. With our confidence the highest it has been all trail, we may have taken the twelve miles a bit lightly. Carrying no packs ensured our motivation remained high with the only other alternative being frozen into human popsicles come nightfall.

***

October 1 - Day 154

There is a frozen pizza cooking in the oven, filling the kitchen with the tantalizing smell of melted cheese and sizzling pepperoni. Six beers cool in the fridge. A map of Glacier National Park is spread on the table, accompanied by sheets of paper filled with scribbled notes and calculations. Sitting around all day is not healthy for any human, but it is certainly not healthy for thru-hikers.

After spending the day brainstorming possibilities, sharing ideas, and speaking our desire to finish the hike, Koozie and I decide to get all logistics down on paper. During our most recent conversation, we were both moved to tears expressing how important hiking this trail is and what it means for us. Working for 5 months toward this goal, only to be halted 75 miles from the finish, is an insult to the previous 2,460 miles hiked and every sacrifice made to get to this point. Our determination is not to be doubted, but our finish-vision can easily get us into trouble that would be better to avoid.

We decide to look at the map, identify all passes, road crossings, and bail-out points, while considering avalanche danger and how to navigate a park covered in three feet of snow. Koozie has to return to work in a week, so his desire to finish this hike is far more desperate than mine. Not having a deadline myself, and knowing I could return in a couple of weeks, if and when the snow melts, I attempt to maintain a realistic outlook on our situation.

Finishing the hike is ideal. Freezing to death in an untimely winter storm is not. I agree to do the job of looking at the map and planning out these last seventy-five miles with him, partly because if the skies remarkably clear up, we need to have a plan and know as much as we can.

One option is to walk the road from here to the border. This is not how either of us imagined the end of our hike, but it is an option. This would ensure we do the seventy-five miles in two or three days, walking as much as possible each day, just to get it done. But we are not out here just to get it done. I did not walk from Mexico to the most spectacular park in the country, just to “get it done” on a paved road. Though roadwalks have had their moments during this journey, I prefer not to use them as a detour around Glacier National Park.

We figure our best option is to get a ride to the Canadian border, either from a local or the Babb Cab, and then walk south through the park back to East Glacier Village. The road to the border is bound to close soon because they do not tend to it once the snow falls. The gate is five miles from the border, so if Park Service does decide to close the road before we get through, it will add another five miles to our hike. This is not much of an addition yet still something we have to consider and make note of.

Every hour I check the weather. Multiple sources and locations are pulled up on my phone. Temperature predictions for town are likely to be twice as cold at elevation, and any forecast for snow is bound to be doubled or tripled in the backcountry and over mountain passes. There is a chance of snow tonight, but only a few inches are predicted at the moment.

Seventy-five miles could be hiked in two days if we really book it. Then again, it took us a full hour to go just one mile the day prior. I would not even be thinking about doing this if I were alone. It is nice to have a friend here to share this experience with. If we did not hike those miles together yesterday, I would have turned around once the visibility went. This is no weather to be going out alone, and being with a buddy is preferred in any dangerous adventure.

That day hike gave us an idea of what to expect at varying elevations and how challenging it is to travel over passes covered with feet of snow in whiteout conditions. If it clears out, at least we will be able to discern the sky from the ridgeline. If it doesn't, we will be navigating every pass by staring at our phones, which could easily lead to a fatal misstep. This is not a situation to take lightly. We could die out there. As we sit here overwhelmed by emotion and ideas, we must appreciate the difficulty of the task before us.

The timer for the pizza goes off. I pull the pie out of the oven and pour two beers before returning to the table to check the weather again. Bad news.

"There is a winter storm warning for the Glacier slash northern Montana area," I say aloud to Koozie.

Our eyes meet. Silence.

"Are you serious?" he replies, hardly attempting to mask the devastation in his eyes and voice.

"Expected accumulation of eighteen to twenty-two inches at Logan Pass," I continue.

Shit.

It contains all the usual warnings: don't travel if you don't have to, the safest place to be is indoors, poor visibility, high winds, etc. Logan Pass is at 6,600 feet, and the pass we went over yesterday topped out at 7,400 feet. Though we are not going over Logan Pass, it is a good measure for other passes and the more remote areas of the park. What was a chance of snow has now become definite snow and lots of it. That's up to another couple feet on top of the two to three already out there.

We sit back and process this news. I read the warning through multiple times, trying to find a loophole and attempting to sure my reasoning that this warning doesn't apply to two experienced thru-hikers on the Continental Divide Trail. I don't listen to REI when they say I shouldn't hike a long trail in trail runners, why should I listen to the National Weather Service when they strongly advise against going outside during a winter storm? We go through our plans one more time, this time accounting for another couple feet of snow.

If it does snow another two feet at elevation, that's it. There is no sense in risking our lives for the completion of a silly hike. Having done all we can in the way of planning, we eat the pizza and try to enjoy our beer. All we can do now is wait for the morning and see how much it snows.

***

October 3 - Day 156

Everyone keeps telling me this train is going to be late. "The train from Glacier to Whitefish is always late," they say. Fine by me. I want to spend as much time in this place as possible. The winter storm dumped a couple of feet in town, and while beautiful to look at from behind the windows of the hostel, it is not safe or sensible to hike through the Glacier backcountry in these conditions. I take another look at the mountains visible from the train station.

"See you in a couple weeks," I tell them.

Though there is no guarantee the sun will return for any amount of time, I say this to reinforce my optimism. The snow came early this year, but locals at the diner admit there is always a chance the sun will return for a week or two toward the end of the month. Possibility of a return remains as I hold on to any amount of hope, wishing for, and visualizing, a sunny week to walk the remaining seventy-five miles through Glacier National Park all by myself during the lonesome, short days of late autumn.

I did not come all this way just to do the last seventy-five miles as quickly as possible in a drowning, life-threatening sea of white. This is supposed to be the best part of the hike. Glacier National Park has been on my to-visit list ever since my brother spent a summer working here as a red bus guide.

The pictures were amazing, and he spoke very highly of it, stimulated by the energy it aroused in him. His eyes light up every time he talks about the area, and I have seen many more people share similar enthusiasm since. I was looking forward to hiking through the backcountry of Glacier. Now that I am here, I'm being turned away.

My daydreams of hiking through Glacier were filled with clear skies and endless views of spectacular mountains. Hiking through the park in these current winter conditions would not be satisfying. Just as with the rest of my hike, I want to enjoy it and take my time. I have enjoyed every day thus far and would not be pushed to do seventy-five miles through feet of snow. This sounds like the opposite of enjoyment.

Either out of stubbornness or confidence, I refuse to admit my journey over. I am taking the train to southern California to see friends, feel the sunshine, and enjoy an unknown amount of zero days. I will keep an eye on the weather for East Glacier and webcams throughout the park. I remain optimistic that enough snow will melt over the next couple of weeks.

The train is not late. In fact, it is early. I climb aboard behind Koozie and take a seat by the window. Unfortunately, he will not be able to finish his hike this year. He has to go back home and return to work. The train starts moving, and I am overwhelmed with sadness. I do not want to go. The idea of me not returning here in a couple of weeks flashes through my mind and lingers longer than expected. There is a chance it will just keep snowing, and I won't be able to finish my hike this year. I have given up too easily. This decision to wait was a foolish one. Nothing else is going through my mind as the snow-covered mountains rush by outside. Within the hour, it is dark outside, and I am left looking at my reflection in the window.

SEVENTY-FIVE MILES

When people ask if I am finished with the hike or how much further I have to go, I tell them, "Seventy-five miles!" Back at Brownie's Hostel in East Glacier Village, seventy-five miles did not seem like much, but the obstacles remaining between me and the Canadian border are far more logistically complicated than walking around a track three hundred times.

I refer to my reason for visiting southern California as a sort of victory lap. It gives me an opportunity to share my incomplete accomplishment with those who encouraged me before I left, supported me throughout, and are now able to prematurely congratulate me. Whether or not I finish the last seventy-five miles is a small detail that will not take anything away from the fact that I spent five months living in the woods, walking from Mexico to (almost) Canada.

From the moment I left East Glacier Village on the Amtrak, I saved the location in my weather app and am checking the forecast every day. The first week out was still cold, but slowly it begins to warm up with less cloudy days and more partly cloudy days predicted. I check the Logan Pass webcam multiple times a day. My heart leaps whenever the sun is out, conversely dropping when the snow is dark gray, protected from melting by thick, dark clouds overhead.

Seventy-five miles. A lot easier said than done, as are most things.

My time in sunny southern California is a mix of experiences and emotions. As friends, we do a great job of listening and conversing, yet there are times I feel out of place. I am the visitor, and thanks to my many months of solitude and walking in silence, my social skills are somewhat lacking. I wear out of company easily and become quiet after a while. I get to see a lot of friends but feel a separation between us.

Our experiences over the past five months have been completely different. While they speak of stressors in life, such as relationships, school, and work, I listen and feel a little sorry for them. Stressors on trail are minimal, which is why I enjoy hiking so much. Unnecessary worries brought on by a desire to pursue monetary purposes are absent for me. Talk of work is dull and uninteresting. Those in school appear helpless: living off loans, drowning in debt, and suffocating under a pile of study guides.

Seventy-five miles. Is that it?

My friends are happy, or at least seem to be. Most of whom I see during my visit are friends I made while living in Mammoth Lakes. They left in the years prior to my departure, moving on to pursue higher education or year-round jobs, eager to be in one place for more than six months without having to move back and forth between towns, lives, and homes. As more people left town to chase other interests and challenges, I was left with growing uncertainty in my own decision to not leave which was less of a decision and more of a continuation of the life I grew to love in a place I began to recognize as home.

I am curious as to what they think of my situation and if I appear happy. I certainly speak very highly of my time on the CDT, but an experience of this sort is difficult to express through words and conversation. The right questions are difficult to ask, and after being asked the same simple queries of, "What was your favorite part?" or, "How many miles do you hike a day?" for the past five months, I desire something more introspective and thought-provoking.

Seventy-five miles. Why didn't I just do the last bit before coming down here?

Perhaps they believe me a little crazy and do not know what to think of their friend ditching them, and the monotony of everyday life. Part of my motivation for hiking was my tiring of constant socialization: asking how each other's day went while catching up on the minuscule details of trivial hours separated. How do we do anything spectacular in a single day? Any worthy accomplishment is the cumulative effort of much time invested working toward a milestone, and this is not usually done in a day or two.

Seventy-five miles. It is easy to forget about the snow in Montana when you are a thousand miles away in sunny California.

Deciding to hike a long trail is no easy decision and possibly the hardest part. Breaking the news to friends and employers is followed by a stream of encouragement and aplomb. If you volunteer to live for six months on trail, there must be something unfulfilling about your life you wish to change or abandon. No hiker is on the trail because they love where their life is at. If you loved where you were in life, you would still be there. We take to the trail because something is missing, suspecting there is more to be discovered and learned.

Seventy-five miles. More like seventy-five and sunny.

The task of completing a thru-hike is certainly attainable because people do it every year. I believe in my ability to accomplish this partly because I have already completed one. It must be attainable, or else what's the use doing it? It must also be difficult, or else what's the use in trying? Participating in any challenge is enjoyable because of the inevitable possibility of failure. Every decision in life is made real and holds consequences because of an ever-looming prospect of failure.

You can fail a difficult test, and you can fail to get to work, but we can do these things properly because of our familiarity with the process. Driving a steel trap at eighty miles an hour down the highway for thirty minutes to get to your workplace sounds pretty dangerous, but with adequate instruction, repetition, and a certain level of focus, this task can be accomplished safely, while navigating around the failure that remains in every blind spot.

Seventy-five miles. The more I say it out loud, the easier it seems.

My decision to thru-hike the Continental Divide Trail was fueled by a desire to accomplish a task for me, removed from the analyzation and judgment of others. Judgment is not bad, and it is not, by any means, good. Judgment just is. Any opinion in reaction to an event, story, or experience is natural and should not be labeled as good or bad. Every action provokes a reaction, and though that reaction may not be voiced, it is still procured through a natural thought process. By no means was I worried about the judgment placed upon my decision by friends or strangers, but even encouragement comes with burdens of expectation and improvement.

Judgment of your actions by those closest to you, whether expressed or not, affects the space and energy existing around your being. I was driven to separate myself from exterior motivations that put me in a certain light, encouraging various aspects of my ego and identity. Though the ego can be useful, it is a shapeshifter, constantly evolving and molding to overcome any obstacle that lies before it. It retains the ability to become an entity of its own, separate from the personal desires and goals I possess, which can be argued to be mine or not as I cannot be completely sure myself.

Seventy-five miles. I keep saying it and laugh a little every time I do. Compared to 2,460 miles, 75 is nothing!

I feel out of place because of my lifestyle for the past five months: on vacation, traveling, no school, no job, no debt, able to live in and enjoy the moment, unafraid of my thoughts and being alone, willing to be uncomfortable, and eager to live with less. This is my first time being around more than two friends at a time since starting this trail. Really good friends. Family. People I have shared a roof with through the winter.

Regardless, after being inside for over an hour, I become agitated and uncomfortable, needing to feel the sunshine on my face and breeze on my body. The walls are confining, and the windows are a poor excuse for outdoor intimacy. The air is dirty in the city, and stale in the houses.

Seventy-five miles. Is that a week of sunshine I see forecasted in Glacier?

XXV.

FLIP-FLOP IN GLACIER

October 18 - Day 171

I return to Glacier National Park on October 18th, dropped off at the border by one of the runners I met a few weeks prior. We take many photos at the border while border patrol and a park ranger hang around to make sure we don't attempt to make a break for it. The only bummer about the final leg of this hike is I have to start at the end. This is so I can walk right back into town and not have to worry about catching a difficult hitch from the border.

This moment of being at the border has been my motivation for the whole summer. The obelisk standing stoically in the clear-cut between the USA and Canada has been the center of my attention while hiking north. The northern terminus does not hold any markings or identification of being the end of such a long and demanding journey, but I am used to the lack of recognition. This trail is all too familiar with the absence of identification.

I wrap my arms tightly around the unforgiving obelisk and speak a few words to the monument, offering my thanks and praise to the silent structure. The stone is cold against my body, and it cools my face. Warm tears fall from my eyes, a combination of joy, sadness, relief, gratitude, and appreciation. Whether it has offered happiness, pain, suffering, or humility, this trail has been my everything for the past six months.

Since leaving the obelisk at Crazy Cook Monument, this northern terminus has been my destination. The thin red line painted on the map has been my course to follow and my path to walk. No matter the emotions I felt, I felt them because of following this track. Any experience I have is at the feet of the trail, and no hug of an inanimate obelisk will be able to demonstrate any portion of my feelings and admiration for this silly footpath.

Glacier National Park is quiet this time of year. One of the downsides of walking through National Parks is they are usually overcrowded with tourists and asphalt rollers. The busy season ended with Labor Day, though stragglers remain through the rest of the month. By mid-October, most of the businesses in the park area have closed for the season and boarded up their buildings in preparation for winter.

Streams gurgle clearly without the distant sound of passing cars. Backcountry campsites are empty and void of visitors, save for the occasional wandering moose. The trail is untracked by people with untouched snow hiding most of the path. The only noise in the forest comes from the consistent crunch of frozen snow beneath my feet.

***

October 19 - Day 172

I wake up in the campground four miles north of Red Gap Pass. It is still dark, but an early wake-up call is necessary. I want to get over the pass before the day warms up, and before the sunlight has a chance to shine on the snow for too long. I pack up by the light of my headlamp and prepare to climb.

Not a quarter-mile from the campground, I come across a river that needs to be forded. There is usually a bridge spanning this channel, but the park service has recently deconstructed all of them in anticipation of winter's snowfall. Wires stretch from abutments on either side with planks stacked neatly beside them. It doesn't look deep, maybe shin height, but the early season snowmelt is sure to be frigid. I knew the bridges would be out but did not know there would be a crossing this soon after leaving camp. I take my shoes and socks off to cross. It is about twenty yards wide, and by the time I get to the other side, my toes are numb.

I get my footwear back on as quickly as possible before continuing up the trail lit by the light wrapped around my head. Within a mile, the trail begins to gain elevation, and the snow becomes deeper. I stop and strap on the snowshoes that I borrowed from the friend who dropped me off at the border. The straps are difficult to adjust, and before long, my fingers are just as cold as my toes.

Walking in snowshoes is slow and tiring. Even with the broad support under my feet, I sink six inches with every step. The initial placement of my foot upon the snow seems strong enough to support my figure, but as I pick up my other foot and transfer weight forward, my standing foot crushes through a thin top layer and sinks down. The process is tedious and tiring.

When hiking over passes covered in snow, it is common practice to start early. Lower nighttime temperatures freeze the snow, theoretically allowing for easier and safer travel. What I thought would be frozen snow throughout is instead a thin frozen top layer with plenty of cold, fluffy stuff remaining underneath. The river crossing only made my feet colder, and with every step into the snow, they get colder still.

I gain elevation slowly but have to stop two miles from the pass to warm up my toes. There is a small, bare patch of ground beside the trail, so I take a break to eat some food and warm my toes up. I take my liner and quilt out of my bag, stuffing the latter into the former to create a well-insulated foot warmer. After removing the snowshoes, along with my frozen footwear, I stuff my feet into the bottom of the bag and pull the liner up past my knees.

This morning I began hiking at 6:15, and it is already 8:15. I have only gone two miles and am making very slow progress toward the pass. Seventeen days off trail provided me plenty of rest, but it also did some damage to my trail legs. Extended time spent at sea level deacclimated my lungs from the thinner mountain air, and I feel as if I am walking through quicksand. The snowshoes weigh my legs down as they collect more snow with every step. I am frustrated, cold, and tired. Welcome back to the CDT, Knots. Seventy-five miles is not going to be a walk in the park if every pass is this difficult.

Thirty minutes later and I am on the side of the trail again. My toes warmed up after some time, but once I started punching through the top layer again, they were cold within ten minutes. I have a couple of layers on the top half of my body, but anything that is not on my person has been stuffed down into the bottom of this liner to help warm my toes and insulate against the cold emanating from the ground and surrounding air. I have not even made it over the first pass of this section, and already I am thinking about the other three passes I have yet to see. What if they are worse than this one? This is the only pass I will hike today, so theoretically, this is the hardest part of my day, the remainder being all downhill to Many Glacier.

I eventually make it over, taking five and a half hours to hike the four miles up to Red Gap Pass. The rest of the miles are much easier in comparison, but my thoughts are dominated by the difficulty of snowshoeing over the pass. I expected my pace to be slower than normal, and while I held no intention of rushing through the last seventy-five miles, I did not expect them to be this difficult and frustrating. I resolve to hitch out once I get to Many Glacier so I can flip back down to Two Medicine and continue north from there. Hiking up the passes from the south side should make the miles more manageable. It is much easier to walk (or slide) down snow-covered passes than it is to walk up them.

***

October 20 - Day 173

The hike up to Red Gap Pass ruined my expectations. Glacier is definitely living up to the hype though this trail still manages to challenge me until my final days. I am humbled as always in the face of Nature. I take one last zero in East Glacier to adjust my backcountry permit and expectations of how far I am able to travel in a day knowing the conditions of the passes and the paths, or lack thereof, leading up to those passes.

I spend a day at Dancing Bears Inn relaxing and eating out of an overpacked food bag, contemplating the end of my hike but mostly enjoying the warmth of being indoors while reading, writing, eating, and being. Zero days in motel rooms are never all that productive, but rest days are as necessary for the mind as they are for the body. Even though I just took two and a half weeks off, I use this day to readjust my mindset and set more realistic goals for the last few days of this hike. The forecast looks favorable for another four days, and I will take advantage of every sunny day I can. The end is near, but I must focus my energy on safely getting through the last fifty-four miles of Glacier National Park.

***

October 21 - Day 174

I return to trail at Two Medicine and begin my approach up the mostly snowless side of Pitamakan Pass. Footsteps lead up to the pass, but I am the first to go any further. The north side of the pass is covered in snow, just as I expected it to be. At the top of Pitamakan Pass, I strap on snowshoes while looking across a snowfield that slopes threateningly toward a cliff.

Thankfully, these snowshoes have spikes on the bottom. Utilizing this feature, I move cautiously across the hard-packed, windblown snow. I am on all fours, hands on the upper side of the mountain, as I shuffle my feet laterally beneath me. The angle of the hill is shallow, but some spots are deeper or slicker than others. Each step is kicked in repeatedly as the snowshoes flop against the sole of my shoes. About 150 feet below me, the cliff drops off another 300 feet to an alpine lake below. I bought a ski pole while in Whitefish in preparation for this winter travel through Glacier, but I rather not test out its self-arresting capabilities. The teeth of the snowshoes hold well and keep me clinging to the side of the mountain.

Carefully and efficiently, I make it across and breathe a great sigh of relief. There is no turning back now. I have already taken as many breaks, zeros, and detours as I possibly can. All I can do is walk and use my best judgment to properly evaluate each situation before carrying on. There are no more footsteps to follow, and the solitude sets in even more. With another obstacle out of the way, I continue north, that much closer to my goal.

I am the only one out here, and this is just the way I like it. When summer was coming to a close, I pictured myself walking through Glacier on an empty trail with yellow leaves holding onto their branches. As autumn started deteriorating into winter, my mental images began to look much whiter, with a thin layer of snow resting upon the surrounding mountains. Now that I am here, it is a combination of both.

Some trees carry brown leaves barely hanging on while others still wear a bold coat of yellow. Remains of life linger, shaking their last shivers before fading into the darkness of winter. A thin layer of snow delicately compliments the landscape in some areas while other spots remain waist-deep. There are no other human footprints, but I do notice a few grizzly and moose tracks. The only wildlife I see throughout the day are the birds that flutter between branches, chirping in the afternoon sun.

***

October 22 - Day 175

I wanted to be on this vacation for as long as possible, and I have certainly pushed the hiking season to its limit of walkability. Hiking north over the passes is much easier than walking south. Trails up to the passes hold some snow but are almost completely melted out. These would have been terribly unsafe to hike in a whiteout. The paths wind along the contour of the mountain with just a few feet between the edge of the trail and a steep drop off. I cringe when I think about the one pass Koozie and I hiked over from Two Medicine to East Glacier Village.

Snow now covers most of the ground, and I am clearly the first one to walk through since the early season snowfall. There is no one out here except for me and the critters of the wilderness. Call me selfish, but I enjoy the silence Nature brings, where the only reminder of civilization is the occasional buzz of an overhead airplane.

On the morning of day 175, I start my hike up to Triple Divide Pass, just underneath Triple Divide Peak. Any water dumped on top of this peak would, theoretically, split three ways down each side of the mountain to end up in three different watersheds: a geographic marvel that seems to be more black magic fuckery than natural brilliance, not that there is much difference between the two anyway.

The closer to the pass I get, the easier the trail becomes as it flattens out and widens with the approach of the saddle. Triple Divide Peak stands tall to my left, a magnificent coming together of three spine-tingling ridges. The morning sun shines heavily on the south side of the mountain, providing safe passage up to the pass. I am amazed the whole way up as I take in the scenery surrounding me. Snow covers the steep and rocky portions of the mountain face that remain in the shade for most of the day. The sun never rises as high as it did the day before, another sacrifice of the season.

A few miles beyond Triple Divide Pass, the trail drops below 5,500 feet, where the snow is completely melted out, and the trail becomes a dry spectacle. Toward the late afternoon, I reach a burn area where it feels like summer once again. I am grateful for another sunny day, probably the last I get to enjoy wearing just shorts and a shirt. Compared to miles walking over snow and through half-melted out trail, I move swiftly through the valley, careful not to walk too fast. This is my shortest day in Glacier National Park, hiking only 11.5 miles from the south side of Triple Divide Pass to the north end of Red Eagle Lake.

Upon reaching the lake and campground, I strip down on the shore and take a dip in the cold water, baptizing myself under the snowmelt of years past. It has been a while since I had the opportunity to go for a swim, and I take advantage of the warmer weather to savor one last soak. I do not bother dressing after my bath and instead lay down in the grass to dry, absorbing the day's final rays as the sun drops toward the tall mountains standing resolute above the lake.

From the winter-teased elevations of Triple Divide Pass to the sun-baked valley where Red Eagle Lake resides, I have gone through four seasons in a single day. This valley is completely void of snow, and while the water is still pretty cold, I warm myself quickly under the forgiving sun. The great ball of light soon moves me into the shadows and casts a vast coolness around the lake. I dress reluctantly before retreating to the campground to set up my tent and prepare dinner.

Two more nights left on trail before I remove myself from where I would rather be. I'm not sure if I'm more concerned about having to leave the trail or if I'm just scared to move on to the next chapter of my life. I much prefer to spend my life watching the world turn than wasting my days staring at a clock and counting the hours for which I am to be financially compensated while waiting for freedom. Why should I invest my time in something that holds indefinite importance, not only in my life span but in the wider scope of humanity and the even wider scope of the universe?

There is nothing. There is no one. I am all there is, and I do not exist. The scenery surrounding me has been here since the beginning. After being carved out by the melting of glaciers, evidence of the last ice age, it all stands stoically before me. Anything I do in this universe is terribly trivial. None of it matters, and though I desire to find purpose in life, the only thing that makes sense to me is spending time in Nature, closer to what is real and closer to that which is greater than me.

Spending every day walking and working toward a goal has held meaning for me. It has given me something to do with the days of the year. This lifestyle allows me to live outside and grow more intimate with anything that may matter. All over the world, we are changing the basic building blocks of life in an attempt to ensure humanity's longevity. Technological advancements and government agendas are determining the progress and condition of human life. Such intentions should not be the objective of society. If anything matters, it is the things that have been here for millions of years and are able to live off no more than what Nature provides.

XXVI.

THE WORST DAY ON TRAIL

October 24 - Day 177

I have been counting down the miles for a while now, and soon I will be counting down the hours. The minutes. The steps. This is my last day on trail and the 177th day since Nick and I began walking from the southern terminus, always getting further from Mexico and closer to Canada. The highs, the lows, and everything between have provided me with great motivation to live life the best I can. My semester of hiking is nearly over, and the lessons I have been taught on trail will certainly apply to my other life.

Over these past few days, tears have been shed at the border, in the motel room, on trail, and at camp. Grateful tears, joyous tears, and sad tears were all spent. I took time to reminisce over the good moments and remembered the bad.

Regardless of what happened, I was always glad to be on trail, where every day is the best day. It has been a challenging yet rewarding six months, during which I have learned a lot about myself and seen some of the more beautiful and remote parts of our country.

The final pass before reaching Many Glacier is difficult, but at this point, I know I can do anything. The clouds have returned, and the wind deafens my eardrums as I am blown sideways, forced down from the pass unceremoniously. As I get closer to the road, I feel myself walking slower, breathing deeper, and seeing more. During the last few miles, I speak words of gratitude to the trees and thank the trail for all it has done. I kiss every last CDT marker I see and wrap my arms around huggable trees. More tears fall from my eyes and drip down my cheek pressed up against the cold bark. I do not want to let go. When I let go, I must continue walking and to keep walking means to get even closer to the end.

On a day where I should be the happiest, I am the most conflicted. The completion of this hike is not measurable against the 176 days that preceded this final one. Personal motivation has brought me here, and while completing this goal gives me reason to be proud, I must find new beginnings elsewhere. Though the end of a hike is the worst part, it has to close at some point. A thru-hike is composed of infinite incredible moments leading up to one final terrible moment of finishing. The joy comes from the doing and is not necessarily dependent upon the completion of.

The accomplishment of finishing a hike comes from the prior days spent walking to reach that point. What makes the final day so rewarding and fulfilling is the time you spent building up to that climax. It is a celebration of the effort and energy invested in such an all-consuming endeavor. I have not been looking forward to ending my hike because it means I have to stop living out of a backpack and sleeping outside. I have run out of trail for now and must move on to prune pastures that may not be as green.

After living on-the-go and out of a backpack for six months, it is time to return to the structure of society. While thru-hiking is a great way to spend six months, I want the activity to aid and improve other areas of my life. Thru-hiking is never just that, and there are plenty of underlying teachings that may not be completely obvious at first. There is never one momentous "Aha!" moment on trail, but there are many smaller moments that hold their own realizations and perception shifts. I return to the other world with new goals, a fresh perspective, and countless memories floating through my head that will have me smiling suggestively as I work through the monotony of the day.

Going on a thru-hike is not going to solve all your problems, if any. Hikers more stubborn than I have lived out of a backpack for six months and not changed a bit. There is no single answer to your purpose in life, just as there is no culmination of sudden enlightenment upon reaching the northern terminus. The border is just another reminder that all things have an end. No matter how enjoyable or awful a thing is, it will not last forever, and the end is always getting closer.

Hiking 2,535 miles from Mexico to Canada has me feeling like a superhuman. I really can do anything I want and actually believe it too. At some point during the hike, I was granted the realization that anything is truly possible. All it takes to accomplish the thing you want to accomplish is time and effort. Whatever it is you want to do, you can. The only thing holding you back is you.

When hiking alone, there is no one around at the end of the day to tell you, "Good job!" or, "You're doing great!" Nobody was watching me hike, and no one was around to cheer me on when the miles were tough. No other voices existed apart from those inside my head or the one I used to speak aloud just to hear somebody talk. Friends and family stood-by for town updates, social media posts, and the occasional phone call, but I did not have to do any of that. The intermittent validation I got from social media posts or the occasional text message was secondary to the motivation I had to provide myself. I rarely had service in the backcountry, and during my worst moments of despair or frustration, there was no one to double-tap me to offer a heart and no one behind a screen, typing words of encouragement.

Though I started this hike with my brother, I was soon enough on my own anyway. I was alone more time than not and had only myself to get along with. If you are doing a hike for someone else, then why are you doing it? At the end of the day, you have to set up camp and eat a cold dinner out of a food bag. You have to walk twenty miles a day through the spectacular and less than spectacular only to do it all again the next day. You hold the world in your hands, and you decide how to wrap it up and make it spin.

Hiking and camping alone for the majority of this journey has allowed me time to think and not to think, letting the moment be while my legs carried me forward only as fast as was ever necessary. Countless times during this journey, I would stop, take a deep breath, look around me, and smile with my arms wide open, laughing to myself with no one around, constantly amazed by the beauty and always appreciative of the opportunity to spend six months hiking the Continental Divide. When life is simple, life is good. Is there anything gooder?

There is no place I want to go because there is nowhere I rather be. This is life. This is living. This keeps me young. It makes me wiser and more patient, grants confidence, which in turn offers humility. Nature gives off a buzz of energy. Trees hum with it, dirt emanates it, and streams constantly shift because of it. This has been the most enjoyable and most challenging six months of my life. The Continental Divide Trail has ruined me when I thought myself already ruined.

The dark clouds that have been hanging above me all day finally break as they reflect the moisture dripping out of my own eyes. I turn off the trail and walk a gravel road into Many Glacier as raindrops fall gently around me. There is a construction crew tending to the lodge and boarding up windows. I wave and try to catch their eyes, but no one bothers to notice. I walk by, take a few pictures on the lakeshore, and rest one last time in an attempt to take it all in. After a while, I shoulder my backpack and continue toward the road. An RV approaches as I reluctantly stick out my thumb. They pick me up and take me into town.

ACKNOWLEDGMENTS

While I walked every mile of trail with my own two feet, I was never alone.

First, I have to thank my parents for their continued support in every facet of my life. Society prefers to shove its youth into a misshapen mold, spilling off-color jello onto the counter of conformity, yet you two continue to encourage my pursuit of an exceptionally abnormal lifestyle. Thank you for showing me the joys of camping, hiking, and being outdoors.

Thanks to my brother for getting me back on trail and being my adventure buddy since day one. Although it's not easy to keep up with your pace, it is always easier to go somewhere when I get to follow you.

Thanks to those who read early drafts of this book, but especially to Riley and Victoria. Your comments, edits, and suggestions helped this book on its way to readable completion. I appreciate the time you both took to read my words and am beyond grateful.

Thanks to everyone who aided me in my efforts along the way and to those who went out of their way to ask if I needed water, food, a place to rest for the night, or a ride somewhere. When our eardrums are being constantly berated with the faults of humanity, it is a great relief to be reassured that we are far from helpless. Thanks to those who were willing to talk to me and share conversation in town or on trail. I learned as much from you as you did from me. While there are no stupid questions, there are certainly repetitive ones.

Continental Divide Trail

ABOUT THE AUTHOR

Brian was born in Virginia in 1992 and has since lived in California, Wisconsin, and Montana. When he is not living out of a backpack, Brian spends his time walking, writing, watching soccer, and working seasonal jobs to save up for the next trip. This is his first book. Visit Brian's website at terminallycurio.us to read more of his words and stay updated on current adventures.

Northern Terminus at the USA/Canada border